ALTERNATIVE TECHNOLOGIES TO REPLACE ANTIPERSONNEL LANDMINES

Committee on Alter
Antipe

Commission on Eng
Office of
Nationa

NATIONAL ACADEMY PRESS
Washington, D.C.

NATIONAL ACADEMY PRESS **2101 Constitution Avenue, N.W.** **Washington, D.C. 20418**

NOTICE: The project that is the subject of this report was approved by the Governing Board of the National Research Council, whose members are drawn from the councils of the National Academy of Sciences, the National Academy of Engineering, and the Institute of Medicine. The members of the committee responsible for the report were chosen for their special competences and with regard for appropriate balance.

This study was supported by Contract/Grant No. V101(93)P-1637, TO#16 between the National Academy of Sciences and the Department of Defense. Any opinions, findings, conclusions, or recommendations expressed in this publication are those of the author(s) and do not necessarily reflect the views of the organizations or agencies that provided support for the project.

International Standard Book Number: 0-309-07349-9
Library of Congress Catalog Card Number: 2001088182

Limited copies of this report are available from:
Division of Military Science and Technology
National Research Council
2101 Constitution Avenue, N.W.
Washington, D.C. 20418

Additional copies are available from National Academy Press, 2101 Constitution Avenue, N.W., Lockbox 285, Washington, D.C. 20055; (800) 624-6242 or (202) 334-3313 (in the Washington metropolitan area); Internet, *http://www.nap.edu*

Printed in the United States of America

THE NATIONAL ACADEMIES

National Academy of Sciences
National Academy of Engineering
Institute of Medicine
National Research Council

The **National Academy of Sciences** is a private, nonprofit, self-perpetuating society of distinguished scholars engaged in scientific and engineering research, dedicated to the furtherance of science and technology and to their use for the general welfare. Upon the authority of the charter granted to it by the Congress in 1863, the Academy has a mandate that requires it to advise the federal government on scientific and technical matters. Dr. Bruce M. Alberts is president of the National Academy of Sciences.

The **National Academy of Engineering** was established in 1964, under the charter of the National Academy of Sciences, as a parallel organization of outstanding engineers. It is autonomous in its administration and in the selection of its members, sharing with the National Academy of Sciences the responsibility for advising the federal government. The National Academy of Engineering also sponsors engineering programs aimed at meeting national needs, encourages education and research, and recognizes the superior achievements of engineers. Dr. William A. Wulf is president of the National Academy of Engineering.

The **Institute of Medicine** was established in 1970 by the National Academy of Sciences to secure the services of eminent members of appropriate professions in the examination of policy matters pertaining to the health of the public. The Institute acts under the responsibility given to the National Academy of Sciences by its congressional charter to be an adviser to the federal government and, upon its own initiative, to identify issues of medical care, research, and education. Dr. Kenneth I. Shine is president of the Institute of Medicine.

The **National Research Council** was organized by the National Academy of Sciences in 1916 to associate the broad community of science and technology with the Academy's purposes of furthering knowledge and advising the federal government. Functioning in accordance with general policies determined by the Academy, the Council has become the principal operating agency of both the National Academy of Sciences and the National Academy of Engineering in providing services to the government, the public, and the scientific and engineering communities. The Council is administered jointly by both Academies and the Institute of Medicine. Dr. Bruce M. Alberts and Dr. William A. Wulf are chairman and vice chairman, respectively, of the National Research Council.

COMMITTEE ON ALTERNATIVE TECHNOLOGIES TO REPLACE ANTIPERSONNEL LANDMINES

GEORGE BUGLIARELLO (NAE), *chair,* Polytechnic University, Brooklyn, New York
H. NORMAN ABRAMSON (NAE), Southwest Research Institute, San Antonio, Texas
THOMAS F. HAFER, Science and Technology Associates, Inc., Arlington, Virginia
J. JEROME HOLTON, Defense Group, Inc., Alexandria, Virginia
LEE M. HUNT, Consultant, Alexandria, Virginia
RICHARD H. JOHNSON, U.S. Army (retired), Alexandria, Virginia
K. SHARVAN KUMAR, Brown University, Providence, Rhode Island
RONALD F. LEHMAN II, Lawrence Livermore National Laboratory, Livermore, California
LARRY G. LEHOWICZ, U.S. Army (retired), Quantum Research, International, Arlington, Virginia
ALAN M. LOVELACE (NAE), General Dynamics Corporation (retired), La Jolla, California
HARVEY M. SAPOLSKY, Massachusetts Institute of Technology, Cambridge
DANIEL R. SCHROEDER, U.S. Army (retired), Vass, North Carolina
MARION W. SCOTT, Sandia National Laboratories, Albuquerque, New Mexico
K. ANNE STREET, Riverside Consulting Group, Inc., Alexandria, Virginia
PATRICK H. WINSTON, Massachusetts Institute of Technology, Cambridge

National Research Council Staff

BRUCE A. BRAUN, Director, Division of Military Science and Technology
JO L. HUSBANDS, Senior Staff Officer, Office of International Affairs
MARGARET N. NOVACK, Study Director
LOIS E. PETERSON, Program Officer
WILLIAM E. CAMPBELL, Administrative Coordinator
CHRISTINA B. MAIERS, Program Specialist (until August 2000)
GWEN ROBY, Senior Project Assistant

Liaisons

Board on Army Science and Technology
GEORGE T. SINGLEY III, Hicks and Associates, Inc., McLean, Virginia

Air Force Science and Technology Board
ANTHONY J. BURSHNICK, U.S. Air Force (retired), Consultant, Springfield, Virginia

COMMISSION ON ENGINEERING AND TECHNICAL SYSTEMS

W. DALE COMPTON (NAE) *chair,* Purdue University, West Lafayette, Indiana
ELEANOR BAUM, Cooper Union for the Advancement of Science and Art, New York, New York
RUTH M. DAVIS (NAE), Pymatuning Group, Inc., Alexandria, Virginia
HENRY J. HATCH (NAE), U.S. Army (retired), Reston, Virginia
STUART L. KNOOP, Oudens and Knoop, Architects, PC, Chevy Chase, Maryland
NANCY G. LEVESON (NAE), Massachusetts Institute of Technology, Cambridge
CORA B. MARRETT, University of Massachusetts, Amherst
ROBERT M. NEREM (NAE), Georgia Institute of Technology, Atlanta
LAWRENCE T. PAPAY (NAE), Science Applications International Corporation, McLean, Virginia
BRADFORD W. PARKINSON (NAE), Stanford University, Stanford, California
BARRY M. TROST (NAS), Stanford University, Stanford, California
JAMES C. WILLIAMS (NAE), GE Aircraft Engines, Cincinnati, Ohio
RONALD W. YATES, U.S. Air Force (retired), Monument, Colorado

Staff

DOUGLAS BAUER, Executive Director
DENNIS CHAMOT, Deputy Executive Director
SYLVIA GILBERT, Administrative Associate
CARLA PAGE, Administrative Assistant
SHARON SEGAL, Financial Officer
CAROL R. ARENBERG, Editor

NOTE: This study was initiated under the auspices of the Commission on Engineering and Technical Systems, which was subsumed in January 2001 by the newly established Division on Engineering and Physical Sciences.

Preface

This National Research Council (NRC) study, commissioned by the U.S. Department of Defense (DOD) in response to a mandate from Congress, addresses the question of whether there are alternatives to antipersonnel landmines (APL)—including technologies, tactics, and operational concepts. The study was conducted at an interesting historical juncture, when the United States is at peace and, at the same time, the number of new technologies rich in military possibilities is unprecedented. The convergence of these two factors presents the U.S. Armed Forces with a unique window of opportunity to develop new systems and concepts to address future challenges.

This is also a moment of heightened international concern about the thousands of civilian casualties that occur every year when APL that have been left in the field after a conflict explode automatically on contact. When military operations are conducted in the midst of a civilian population, the problem is compounded because today's mines cannot discriminate between friend and foe, belligerent and civilian. It is important to note, however, that APL fielded by U.S. forces, except for APL in storage in Korea, are designed to self-destruct or self-deactivate at a preset time. Therefore, they do not remain a danger indefinitely.

No simple device today can provide capabilities comparable to those of APL, both as self-standing devices and as a part of other systems. Devices currently under development include mine-like devices that do not explode automatically on contact and nonlethal devices that could complement lethal devices and systems. Thus, the functions of today's APL could be performed by a combination of devices, carefully planned tactics, and appropriate operational procedures. In some circumstances, however, replacing APL could lead to higher casualties to our ground forces and/or could reduce our military capabilities.

The committee believes strongly that the development of new systems with decoupled sensing, communication, and explosive functions and the creation of networks of technologically sophisticated tactical sensors would greatly increase the situational awareness and power of war fighters and help meet the goal of ensuring the information superiority of U.S. forces. These systems would also respond to the humanitarian principle manifested in the Ottawa Convention of eliminating antipersonnel devices that explode on contact. Although these new systems are bound to have vulnerabilities different from those of APL, these vulnerabilities could be greatly reduced by the application of appropriate technologies. Therefore, DOD should move rapidly to support pertinent research and development to create fieldable systems.

The NRC committee that produced this report worked diligently in the limited time available to respond to DOD's request. The report draws on presentations to the committee in both public and closed sessions by representatives of government, industry, and nongovernmental organizations, interviews, research by committee members, and their expertise and judgment.

The committee is grateful to everyone who contributed to the study, particularly Margaret Novack, study director, and Lois Peterson, program officer, who worked tirelessly to see the study through to completion.

George Bugliarello, Chair
Committee on Alternative Technologies
to Replace Antipersonnel Landmines

Acknowledgments

The study was conducted under the codirectorship of two National Research Council commission-level offices: the Commission on Engineering and Technical Systems (CETS) and the Office of International Affairs (OIA). An oversight group was formed to ensure unity of effort and to provide an internal review of this report. We wish to thank the following individuals for their participation in the oversight group:

Henry J. Hatch (NAE), *chair*, U.S. Army (retired)
John Baldeschwieler (NAE), California Institute of Technology
Nicole Ball, University of Maryland
Ruth M. Davis (NAE), Pymatuning Group, Inc.

This report has been reviewed in draft form by individuals chosen for their diverse perspectives and technical expertise, in accordance with procedures approved by the NRC's Report Review Committee. The purpose of this independent review is to provide candid and critical comments that will assist the institution in making its published report as sound as possible and to ensure that the report meets institutional standards for objectivity, evidence, and responsiveness to the study charge. The review comments and draft manuscript remain confidential to protect the integrity of the deliberative process. We wish to thank the following individuals for their review of this report:

Arden L. Bement (NAE), Purdue University
John Christie, Logistics Management Institute
Stephen D. Goose, Human Rights Watch
Jerome H. Granrud, U.S. Army (retired)
Thomas McNaugher, RAND Corporation
Hyla Napadensky (NAE), Napadensky Energetics
Richard I. Neal, U.S. Marine Corps (retired)
Francis B. Paca, VSE Corporation
William C. Schneck, U.S. Army Night Vision and Electronic Sensors Directorate
Sarah Sewall, Carr Center for Human Rights
John F. Troxell, U.S. Army War College
Gerold Yonas, Sandia National Laboratories

Although the reviewers listed above have provided many constructive comments and suggestions, they were not asked to endorse the conclusions or recommendations, nor did they see the final draft of the report before its release. The review of this report was overseen by Lewis M. Branscomb, NAE, Harvard University. Appointed by the National Research Council, he was responsible for making certain that an independent examination of this report was carried out in accordance with institutional procedures and that all review comments were carefully considered. Responsibility for the final content of this report rests entirely with the authoring committee and the institution.

Contents

Tables, Figures, and Boxes

TABLES

FIGURES

BOXES

Acronyms

ADAM	Area-Denial Artillery Munition
APL	antipersonnel landmines
AT	antitank (mine)
BAT	Brilliant Antiarmor Submunition
BATAAM	BAT Antiarmor Munition
BNLM	Bounding Nonlethal Munition
C4ISR	command, control, communications, computers, intelligence, surveillance, and reconnaissance
CCW	Convention on Conventional Weapons
CFC	Combined Forces Command
CINC	Commander in Chief
CJCS	Chairman of the Joint Chiefs of Staff
CLADS	Canister-Launched Area-Denial System
DARPA	Defense Advanced Research Projects Agency
DOD	U.S. Department of Defense
DPICM	Dual-Purpose Improved Conventional Munition
DWSC	Distributed Web Sensor Complex
EWSS	Early Warning Subsystem
FASCAM	Family of Scatterable Mines
FCS	Future Combat System
FM	field manual
GEMSS	Ground-Emplaced Mine-Scattering System
GPS	Global Positioning System
HESF	Hand-Emplaced Sensor Field
HMMWV	High-Mobility Multipurpose Wheeled Vehicle
JSTARS	Joint Surveillance Target Attack Radar System
LADAR	laser radar
LDMG	LADAR-directed machine gun
MEMS	microelectromechanical system
MLRS	Multiple-Launch Rocket System
MOPMS	Modular-Pack Mine System
MOSAIC	multifunctional, on-the-move, secure, adaptive, integrated communication
NATO	North Atlantic Treaty Organization
NSD-A	nonself-destructing (antipersonnel landmine) alternative
ORD	Operational Requirements Document
PIP	Product Improvement Program
RAAMS	Remote Antiarmor Mine System
RAAMS-NL	Remote Antiarmor Mine System-Nonlethal
RADAM	Remote Area-Denial Artillery Munition
RD-Sensor	RAAMS enhanced with telemetry and sensor package
RD-Telemetry	RAAMS enhanced with telemetry
RD-WAM	Remotely Delivered Wide Area Munition
RRASMS	Radio/Radar Sensor Munition System
SADARM	Sense and Destroy Armor Munition
SSDMS	Small Short-Duration Mine System
SFW	Sensor Fuzed Weapon
SINCGARS	Single-Channel Ground-to-Air Radio System
TACMS	Tactical Missile System
UNC	United Nations Command
URAS	Unmanned Remote Ambush System
WAM	Wide Area Munition

Executive Summary

BACKGROUND

Military forces use landmines or landmine-like devices because they are capable of autonomously delaying or killing the enemy at a safe distance from friendly forces. Landmines fall into two broad categories. Antipersonnel landmines (APL) are intended to kill or disable soldiers on foot (a dismounted force).[1] Antitank landmines (AT mines) are used against vehicles, such as tanks and armored personnel carriers (a mounted force). Mixed systems, which combine both AT mines and APL in the same munition, are typically used against an enemy force that is mostly mounted but is accompanied by significant numbers of dismounted soldiers. APL in mixed systems are intended to prevent or discourage foot soldiers from penetrating or breaching an AT minefield.

Landmines are essentially tactical and operational weapons, although on occasion they also have strategic implications. When used tactically, landmines are usually employed during battlefield engagements of relatively limited duration to disrupt an enemy's progress. There are also long-term strategic landmine missions, such as border protection, as is the case in Korea.

Minefields are used to place an enemy in a vulnerable position that can be exploited by friendly forces, cause the enemy forces to divide, interfere with enemy command and control, inflict damage on enemy personnel and equipment, exploit the capabilities of other weapon systems by delaying enemy forces in an engagement area, and protect friendly forces from enemy infiltration. U.S. doctrine requires that minefields be mapped, marked, and eventually cleared. Despite these precautions and for a variety of reasons, U.S. landmines also occasionally kill friendly personnel, typically in hastily marked minefields. In addition, if the tide of battle changes rapidly, U.S. mines previously emplaced during defensive missions could become an obstacle to the execution of rapid offensive maneuvers.

Over time, landmines used by the United States and other countries with advanced military forces have become more complex, more effective, and easier to use. Advances in the 1970s also led to the development of landmines capable of destroying or deactivating themselves after a given time. All landmines currently in U.S. stocks, with the exception of those intended for the defense of Korea, are self-destructing and/or self-deactivating.

Other nations and nonstate actors, unfortunately, have resorted to less technologically advanced landmines, which are inexpensive, easily obtainable, and highly effective. Most of these simple, nonself-destructing mines are deployed with no thought to keeping track of their locations. As a result, millions of these devices are still strewn across old battlefields. APL have killed or maimed thousands of innocent civilians in the last 25 years and impeded the restoration of normal activities after conflicts have ended. Humanitarian groups, international organizations, and many governments around the world have increasingly identified these residual hazards as a threat to innocents and demanded that all APL be eliminated.

Protocol II of the 1980 *Convention on Prohibitions or Restrictions on the Use of Certain Conventional Weapons Which May Be Deemed to Be Excessively Injurious or to Have Indiscriminate Effects* (otherwise known as the Convention on Conventional Weapons, or CCW) was the first international treaty to attempt to regulate the use of landmines. A review of the CCW in 1996 led to Amended Protocol II, which, among other things, distinguished between the use of APL and AT mines and restricted the uses of all APL. The United States has signed and ratified the CCW, including Amended Protocol II.

Subsequent national and international campaigns demanding a total ban on APL led to the *Convention on the*

[1] The Convention on Conventional Weapons, Amended Protocol II, defines an APL as "a mine primarily designed to be exploded by the presence, proximity, or contact of a person and that will incapacitate, injure, or kill one or more persons" (see Appendix E).

Prohibition of the Use, Stockpiling, Production and Transfer of Anti-Personnel Mines and on Their Destruction (the Ottawa Convention), which was signed by 122 countries in Ottawa, Canada, on December 3, 1997, and entered into force in March 1999. As of September 2000, 139 nations had signed the Ottawa Convention, including all NATO member states, except the United States and Turkey, and all European Union member states, except Finland. The Ottawa Convention bans the use of all APL,[2] whether used alone or in mixed systems, including those that are self-destructing and self-deactivating.[3] Signatories are prohibited from developing, producing, acquiring, or stockpiling APL, as well as assisting, encouraging, or inducing anyone else to undertake these actions. All APL currently held by signatories must be destroyed within four years of the signing.

Despite showing early support for a ban on APL and taking the lead in efforts to ameliorate residual effects, the United States did not sign the Ottawa Convention. President Clinton stated that the United States would consider acceding to the convention when alternative technologies that provide capabilities similar to those of APL have been identified and fielded. He also announced that the United States would undertake an active research and development program to find such alternatives. At the same time he established the presidential policy that after 2003, the United States would no longer use pure APL[4] outside Korea, where landmines are considered particularly important. If alternatives for Korea and for mixed systems can be found by 2006, the president said, the United States will sign the Ottawa Convention. In the meantime, the United States has destroyed three million nonself-destructing mines.

U.S. SEARCH FOR ALTERNATIVES

In 1997, the U.S. Department of Defense (DOD) began the task of developing alternatives to APL. DOD initially adopted a two-track approach. Track I, led by the U.S. Army, was a search for alternatives to the nonself-destructing landmines used in Korea. Track II, a Defense Advanced Research Projects Agency (DARPA) program, was focused on an assessment of long-term, more technologically advanced alternatives that would effectively prevent access to an area. In 1999, Congress provided funds to add a third track. The goal of Track III, which overlaps both Track I and Track II, is to find existing and new technologies and operational concepts that can provide an equivalent to the capabilities of (1) nonself-destructing APL; (2) APL used in mixed AT mine systems; and (3) current mixed landmine systems, including AT mines with antihandling devices.

The National Academies

As part of the Track III initiative, DOD contracted with the National Academy of Sciences to conduct a study of existing and new technologies that might provide an alternative to APL. In response, the Committee on Alternative Technologies to Replace Antipersonnel Landmines was established. The committee was asked to (1) identify and examine possible tactics, technologies, and operational concepts that could provide tactical advantages similar to those provided by APL by 2006; (2) suggest a near-term alternative technology, weapon system, or combination of systems that could be derived from known, available systems or that could provide a short-term solution if the recommended alternative will not be available by 2006; and (3) describe how the identified technologies and systems could be used consistently with current tactical doctrine and operational concepts or recommend changes in tactics or operational concepts. This report is the result of that study.

Political Context for the Study

The committee was asked to consider alternatives that would provide tactical advantages to U.S. forces similar to those provided by APL. The committee also recognized that it had an opportunity to recommend alternatives, especially improved sensors and communications that would be more militarily effective than current APL. However, considering the presidential policy and official statements regarding APL, the committee recognized that one reason for the search for alternatives was to enable the United States to accede to the Ottawa Convention. The committee made no judgment as to whether the United States should accede to the Ottawa Convention.

Conclusion 1. The major reasons for seeking alternatives to current antipersonnel landmines (APL) are humanitarian concerns, compliance with the Ottawa Convention, and enhanced military effectiveness. Indeed, this study would not have been empanelled were it not for the Ottawa Convention. The current inventory of self-destructing and self-deactivating U.S. APL is militarily advantageous and safe. They achieve desired military objectives without endangering U.S. warfighters or noncombatants more than other weapons of war, but they are not compliant with the Ottawa Convention. However, humanitarian concerns and Ottawa compliance are not always synonymous. In fact, some of the apparently Ottawa-compliant alternatives examined by the committee may be less humane than present U.S. self-destructing and self-deactivating landmines.

[2] The convention does not prohibit command-detonated munitions, such as the Claymore, although they are customarily described as APL.

[3] The negotiators did not allow for the inclusion of self-destructing and self-deactivating APL for several reasons. These mines still fit the definition of APL, and no exceptions were to be made. If an exception had been made for these mines, primarily in the inventory of only the United States and a few western European countries, exceptions might have had to be made for weapon systems of other countries.

[4] "Pure" APL are APL used alone and not as part of a mixed system.

Recommendation 1a. If the decision is made to accede to the Ottawa Convention, a transition period may be necessary before implementation to maintain current U.S. military capabilities until suitable alternatives can be made available. During that transition, existing self-destructing and self-deactivating antipersonnel landmines should be retained, both in their stand-alone form and as part of mixed systems.

Recommendation 1b. Of the solutions not compliant with the Ottawa Convention, simply retaining the current self-destructing and self-deactivating mines would be the best course of action.

COMMITTEE ASSESSMENTS

After reviewing the functions performed by landmines, as well as the context within which these functions might be needed now and in the future, the committee sought alternative ways of performing the same tasks. The committee reviewed a broad variety of nonmateriel alternatives, including innovative operational concepts and military tactics, and materiel alternatives, such as weapon systems, nonlethal devices, and improved sensors and communications. The alternatives considered included existing systems, concepts under consideration or development, and new concepts. Although a number of these ideas are not fully developed, they might be a basis for new approaches for the future.

A prominent feature of many alternatives is the "man-in-the-loop," which ensures a positive identification of an intruder before a response element is activated. The man-in-the-loop concept envisions a soldier/operator positioned in such a way that he can observe the minefield and determine whether or not the intruder is a friend, an enemy, or a noncombatant. New technologies, especially improved sensors and communications, would enable the soldier/operator to make a faster, more accurate identification, which would be beneficial for humanitarian purposes and would reduce fratricide. However, a man-in-the-loop also introduces a variety of new vulnerabilities.

Conclusion 2. The rapid emergence of new technologies after 2006 will create opportunities for the development of systems that can outperform today's antipersonnel landmines and that would be compliant with Ottawa.

Recommendation 2a. The development of sensor-net technology should be pursued aggressively and eventually incorporated into a fully militarized, deployed system characterized by networking, strong detection and tracking capabilities, robustness, low power consumption, low cost, covertness, low probability of intercept, easy deployment, and disposability.

Recommendation 2b. Investments already being made in new technologies for other purposes should be leveraged and applied to the search for alternatives to antipersonnel landmines.

Evaluation Methodology

Unfortunately, the committee did not have enough time or resources to use independent modeling or simulations in evaluating the alternatives. Therefore, the committee developed a score sheet to assess systematically the effectiveness of alternatives. The resulting analysis is admittedly subjective, and the results are qualitative rather than quantitative.

Guided by the Statement of Task, the committee first screened alternatives in terms of their availability by or before 2006. Because 2006 is near, especially in terms of the steps required for fielding an alternative system, and because remarkable improvements in technology are forecast for the near future, alternatives that might become available after 2006 were also considered.

Each of the alternatives was considered against a baseline, depending on whether it was intended to be used against dismounted or mounted targets. The M14 and M16, current APL that are nonself-destructing and require hand emplacement, were used as the baseline for alternatives against dismounted targets. The Volcano (M87), a mixed system that includes self-destructing APL to protect AT mines, was used as the baseline for alternatives against mounted targets.

The committee used several criteria to judge the alternatives against these baseline systems: military effectiveness; humanitarian concerns; technical risk; tactics and operational concepts; and cost. To determine whether an alternative would meet DOD's military requirements, the committee used the two mission need statements developed by DOD for APL alternatives as a basis for assessing military effectiveness. The humanitarian intent of international agreements such as the Ottawa Convention and the CCW Amended Protocol II, although not specified in the Statement of Task, was an implicit basis for this study and was also considered a criterion. The committee also considered the overall technical risk of an alternative, that is, whether the technology could feasibly be developed and manufactured. The committee also indicated whether an alternative would require a change in tactics and operational concepts. The last criterion was cost. Although the sponsor indicated that cost should not be a driving factor in the selection of an alternative, the committee decided a rudimentary consideration of cost was necessary.

Table ES-1 lists all of the current systems and alternatives considered by the committee. Descriptions can be found in the body of the report. The alternatives that are mentioned in the committee's conclusions and recommendations are described briefly below.

NONMATERIEL ALTERNATIVES

The committee first considered whether nonmateriel alternatives, such as changes in tactics and operational concepts, could fully compensate for the elimination of APL.

TABLE ES-1 Current and Potential Systems Considered in This Report

System Name	Mine/Nonmine	SD/SDA	Lethal/Nonlethal	Ottawa Compliant**	Dismounted Threat	Mounted Threat	Origin of System
Alternatives Available Now							
Claymore (M18)	APL	N	L	Y	H		E
Volcano (M87A1)	AT	Y	L	Y		R	E
Remote Antiarmor Mine System (RAAMS)	AT	Y	L	Y		R	E
Hornet/Wide Area Munition (WAM)	AT	Y	L	Y		H	E
Maverick (AGM-65)	n/m	N	L	Y		R	E
Longbow Hellfire (AGM-114)	n/m	N	L	Y		R	E
Sensor Fuzed Weapon (SFW)	n/m	N	L	Y		R	E
Sense and Destroy Armor Munition (SADARM)	n/m	N	L	Y		R	E
Brilliant Antiarmor (BAT) Submunition	n/m	N	L	Y		R	E
Alternatives Available by 2006							
Hand-Emplaced Sensor Field (HESF)	n/m	n/a	n/a	Y	H		C
Nonself-Destructing Alternative (NSD-A) **Track I**	APL	Y	L	***	H		D
Sphinx-Moder Perimeter Defense System	APL	N	L	Y	H		D
Multiple-Shot Claymore Mine	APL	N	L	Y	H		C
Bounding Nonlethal Munition (BNLM)	n/m	n/a	N/L	Y	H		D
Taser Nonlethal Munition	n/m	n/a	N/L	Y	H		D
Wide Area Munition Product Improvement Program (WAM PIP)	AT	Y	L	Y		H	D
Remote Area-Denial Artillery Munition (RADAM) **Track I**	Mix	Y	L	N	R	R	D
RAAMS Enhanced with Telemetry (RD Telemetry)	AT	Y	L	Y		R	C
Canister-Launched Area-Denial System (CLADS)****	n/m	n/a	N/L	Y	R	R	D
Volcano-CLADS	Mix	Y	L	Y	R	R	C
AT Pure-Modular-Pack Mine System	AT	Y	L	Y		R	C
AT Pure-Gator	AT	Y	L	Y		R	C
Dual-Purpose Improved Conventional Munition (DPICM) with Random Fuzing (Popcorn)	Mix	Y	L	Y	R	R	D
Small Short-Duration Mine System (SSDMS)	Mix	Y	L	N	R	R	O
Alternatives Available After 2006							
Radio/Radar Sensor Munition System (RRASMS)	APL	Y	L	Y	H		C
Unmanned Remote Ambush System (URAS)	APL	Y	L	Y	H		C
Tags/Minimally Guided Munitions **Track II**	n/m	n/a	n/a	Y	R		D
Laser Radar-Directed Machine Gun (LDMG)	n/m	n/a	L	Y	H		C
Distributed-Sensor Antipersonnel "Minefield"	n/m	n/a	L	Y	R		C
Distributed Web Sensor Complex (DWSC)	n/m	n/a	n/a	Y		R	C
Raptor	AT	Y	L	Y		R	D/C
RAAMS Enhanced with Telemetry and Sensor Package (RD Sensor)	AT	Y	L	Y		R	C
Remotely Delivered Hornet/WAM (RD-WAM)	AT	Y	L	Y		R	C
Self-Healing Minefield **Track II**	AT	Y	L	Y		R	D
BAT Antiarmor Munition (BATAAM)	AT	n/a	L	Y		H	C
Early Warning Subsystem for Remotely Delivered AT Minefields (EWSS)	n/m	n/a	n/a	Y		R	C
RAAMS with Nonlethal Capability (RAAMS-NL)	Mix	Y	N/L	Y		R	C

TABLE ES-1 *Continued*

System Name	Mine/Nonmine	SD/SDA	Lethal/Nonlethal	Ottawa Compliant**	Dismounted Threat	Mounted Threat	Origin of System
Existing Mine Systems*****							
M14*	APL	N	L	N	H		E
M16	APL	N	L	N	H		E
Claymore (M18)	APL	N	L	Y	H		E
Pursuit-Denial Munition (PDM)	APL	Y	L	N	H		E
Ground-Emplaced Mine Scattering System (GEMSS)	Mix	Y	L	N	R	R	E
Modular-Pack Mine System (MOPMS)	Mix	Y	L	N	H	H	E
Area-Denial Artillery Munition (ADAM)	APL	Y	L	N	R		E
Remote Antiarmor Mine System (RAAMS)	AT	Y	L	Y		R	E
Gator	Mix	Y	L	N	R	R	E
Hornet/Wide Area Munition (WAM)	AT	Y	L	Y		H	E
Volcano (M87)	Mix	Y	L	N	R	R	E
Volcano (M87A1)	AT	Y	L	Y		R	E

Key: SD/SDA = self-destructing/self-deactivating, APL = antipersonnel landmine, AT = antitank landmine, mix = combination of APL and AT, n/m = nonmine, n/a = nonapplicable, N = no, Y = yes, L = lethal, N/L = nonlethal, R = remotely delivered, H = hand emplaced, E = existing system, C = committee concept, D = in development, O = other.

*Systems in bold italics would be unavailable if the United States ratifies the Ottawa Convention.

**Compliance is based on the definition in the Ottawa Convention.

***Ottawa compliance would depend on whether the battlefield override switch was part of the design.

****This system is assumed to be used alongside AT mines.

*****Existing mine systems are discussed in Appendix C.

Conclusion 3. By 2006, alternative tactics or operational concepts could not, on their own, provide tactical advantages similar to those provided by antipersonnel landmines, without a significant increase in force structure. In certain situations, however, some nonmateriel alternatives might be useful: increased reconnaissance forward; more soldiers or weapon systems in a given battlefield area; more command-detonated Claymores to protect against a dismounted enemy; antitank mines remotely delivered "just in time" to support a maneuver and inhibit the enemy's ability to breach; and speed, mobility, and offensive tactical operations.

MATERIEL ALTERNATIVES

Alternatives Available Today

Of the five APL currently in the U.S. arsenal, only the Claymore, which is activated by a man-in-the-loop, can be used under the terms of the Ottawa Convention. All three existing AT mines are usable under the Ottawa Convention, but APL munitions could not be used to protect them.

In addition to landmines, several other systems have been proven effective against tanks and large ground vehicles. All of these are air-delivered, precision weapons, however, and probably could not be rapidly delivered on target. Each of these alternatives fell well short of meeting the military effectiveness criteria compared to the Volcano baseline. Although not included in the scoring criteria, the committee was also concerned about the unintended consequences of unexploded ordnance that might result from these weapons. These residual effects could be worse than those of self-destructing and self-deactivating APL.

Alternatives Available by 2006

Between now and 2006, many innovations will be made in weapons technology and sensors and communications. Alternatives that use these technologies will feature new characteristics, such as separation of sensors and kill mechanisms and improved communications between sensors and soldiers. However, unless DOD gives these new technologies a very high priority, six years will not be long enough for the weaponization of any innovative technology.

Alternatives for Use Against Dismounted Targets

The committee evaluated six alternative deterrents to dismounted threats, four of which are described below: the Nonself-Destructing Alternative (NSD-A) Program; the Hand-Emplaced Sensor Field (HESF); the Bounding Nonlethal Munition (BNLM); and the Taser nonlethal munition.

The DOD Track I concept, NSD-A, whose man-in-the-loop design makes the system Ottawa compliant, has a high potential of providing tactical advantages for U.S. forces similar to those provided by current M14 and M16 APL. To be available for implementation by 2006, this system would require concerted, aggressive development and a streamlined acquisition process. Enhancements to the NSD-A system, such as additional sensors and nonlethal elements, could be added over time.

The DOD must also decide whether or not to include a capability in the software design of the NSD-A[5] to permit the soldier/operator to put the system in an autonomous mode. This capability has been referred to as the "battlefield override switch." With this feature engaged, the man-in-the-loop would no longer be necessary to activate the munition, which would become a conventional, target-activated, self-destructing APL. The committee recognized that the NSD-A with the switch would provide greater military flexibility in responding to an intruder. However, the committee concluded that the NSD-A without a battlefield override switch would have significant tactical advantages over the existing M14/M16 APL and would reduce the potential for fratricide and noncombatant casualties.

Conclusion 4. For use against dismounted forces, the Track I alternative to nonself-destructing landmines (NSD-A) could provide, by 2006, similar or enhanced tactical advantages for U.S. forces as compared to those provided by current nonself-destructing antipersonnel landmines. The battlefield override switch, a software capability that allows the system to operate autonomously, is highly contentious because, as presently designed, it would render the NSD-A non-Ottawa compliant. Even though the timing of a decision on the switch or other programmatic delays could jeopardize the timeline, the NSD-A system appears to be technically mature enough to be available by 2006. This weapon system could be greatly enhanced in the future by planning for the inclusion of additional sensors, nonlethal elements, and an Ottawa-compliant battlefield override capability.

[5] A separate study is under way by an office within DOD to assist with this decision.

Recommendation 4a. The development and production of the Track I alternative to nonself-destructing landmines (NSD-A) system should be aggressively pursued to ensure its availability by 2006.

Recommendation 4b. Two suites of weapon software should be developed simultaneously in preparation for a presidential decision concerning the Ottawa Convention. If compliance with the Ottawa Convention were desired, the battlefield override switch, as currently designed, would not be used in the production of the NSD-A. If the president decides that other considerations outweigh Ottawa compliance, the option of retaining the switch would be available. In any case, Ottawa-compliant variations to the battlefield override switch should be explored to provide the United States with greater flexibility.

The HESF could exploit the effectiveness of current weapons by providing early warning and enabling man-in-the-loop control. The sensor field would be a combination of sensor technologies, including existing military systems, off-the-shelf technologies, and sensors being actively developed by the military science and technology community. The operator and his chain of command would respond to confirmed enemy targets with an appropriate kill mechanism.

Recommendation 4c. Sensor technology should be leveraged immediately to develop sensor systems to improve a soldier's ability to discriminate among friends, foes, and noncombatants in all terrain and all weather conditions at much greater battlefield ranges.

Two promising nonlethal alternatives, the BNLM and the Taser nonlethal munition, were also considered as deterrents to dismounted threats. Both weapons could be developed eventually as remotely delivered devices to provide protection against dismounted breaches of AT minefields. Nonlethal alternatives are described and assessed later in this summary.

Alternatives for Use Against Mounted Targets

The committee compared nine alternatives to the Volcano M87 baseline, four of which are discussed below: the Remote Antiarmor Mine System (RAAMS) enhanced with telemetry (RD-Telemetry); the Hornet/Wide Area Munition (WAM) Product Improvement Program (PIP), the WAM PIP; the Remote Area-Denial Artillery Munition (RADAM); and the Canister-Launched Area-Denial System (CLADS). The committee was also provided with descriptions of systems under consideration by DOD as part of the Track III search for alternatives. None of these systems had been developed enough to be assessed, although several did appear to be promising. Because of the need to protect proprietary information, none of them is described here.

One concept developed by the committee was the RD-Telemetry (RAAMS enhanced with telemetry). This concept would involve upgrading the existing RAAMS projectile, which contains AT mines, with a subminiature telemetry and communications package that could calculate the precise locations of dispensed mines and send the information back to friendly forces. Although significant research and development would be necessary, the technology might be useful not only for RD-Telemetry, but also for other submunitions.

The system the committee considered the best alternative against a mounted enemy that might be available by 2006 was the Hornet/WAM PIP. This two-phased, evolutionary improvement program for the existing Hornet/WAM would add a man-in-the-loop to control the minefield, better sensors to improve target detection, and an improved dual-purpose warhead. The WAM PIP's much greater kill-radius would provide military advantages over the baseline Volcano M87. The disadvantages of this mine are its large size and that it cannot be remotely delivered.

Conclusion 5. Under current policy, no fully equivalent alternative to mixed systems is likely to be available by 2006. Other than the Track III search for an alternative, little is being done that could lead to the fielding of a satisfactory alternative. The Hornet/Wide Area Munition (WAM), with its large lethal radius and antihandling device, could replace most of the tactical functions currently provided by mixed systems but has no remote delivery capability. If a satisfactory remote delivery capability could be developed by 2006, the Hornet/WAM appears capable of performing the mixed-minefield mission satisfactorily.

Recommendation 5a. Promising Track III concepts should be developed into weapon system programs. The development of any of these concepts by the 2006 deadline, however, would require that considerable additional resources be allocated for development and procurement.

Recommendation 5b. The feasibility, cost, and schedule of providing a remote delivery option for the Hornet/Wide Area Munition should be investigated. Shock hardening of the mine to withstand the impact of remote delivery appears to be an Ottawa-compliant, low-risk solution to current mixed minefields.

One mixed system considered by the committee was RADAM (Remote Area-Denial Artillery Munition), a concept under development by the DOD Track I. The RADAM would combine existing Remote Antiarmor Mine System (RAAMS) AT mines and the Area-Denial Artillery Munition (ADAM) APL, which are now fired separately, into one projectile. This would necessarily reduce the number of AT mines per projectile, so more projectiles might be required to cover a given area. Although APL in mixed systems are acceptable under current presidential policy, they would not comply with the Ottawa Convention. Until another alternative is developed, using ADAM and RAAMS together, rather than developing RADAM, would be a better way to maintain the mixed capability of artillery-delivered scatterable mines.

Conclusion 6. The Remote Area-Denial Artillery Munition (RADAM), a mixed system, provides little or no military advantage over the combined use of the Remote Antiarmor Mine System (RAAMS) and the Area-Denial Artillery Munition (ADAM). Because RADAM would be no more compliant with the Ottawa Convention than the ADAM/RAAMS combination, funding for its development could be better spent on accelerating the development of an Ottawa-compliant alternative. If DOD determines that an artillery-delivered mixed system must be maintained, there are two options: (1) request a change in presidential policy to allow the continued use of ADAM to be fired in tandem with RAAMS; or (2) develop RADAM. The latter option would require taking the Ottawa-compliant RAAMS out of the inventory to create a new non-compliant munition.

Recommendation 6. Until a long-term solution can be developed, the Area-Denial Artillery Munition (ADAM) should be retained in the inventory for use with the Remote Antiarmor Mine System (RAAMS). Production of the Remote Area-Denial Artillery Munition (RADAM) should be halted and funding redirected toward the development of long-term alternatives for mixed systems.

The nonlethal CLADS, a joint Army-Marine Corps nonlethal program currently on hold, was evaluated both as a weapon launched separately from AT mines and as part of a mixed system in the same canister as Volcano AT mines. CLADS emits an audible warning signal and projects rubber balls when activated by a trip wire. In general, CLADS is a promising, nonlethal APL alternative that may provide some protection for AT minefields from dismounted breaches.

Conclusion 7. Although nonlethal variants by themselves cannot replace antipersonnel landmines, they would be useful in certain military operations. U.S. forces will face a broad range of potential scenarios in the future, from peace operations to intense full combat. With nonlethal variants, U.S. forces could mount a graduated response in situations where the threat is unclear, such as peace operations, or if large noncombatant populations were in the immediate tactical area. Nonlethal weapons have several advantages: they can be used in a broad variety of circumstances; they can be triggered automatically; and they do not require man-in-the-loop operation to be Ottawa compliant, which could improve the timeliness of a response and lessen the burden on the soldier/operator.

Recommendation 7. The development of nonlethal variants to support antipersonnel landmine alternatives should be emphasized. Funding should be restored and development accelerated for the nonlethal Canister-Launched Area-Denial System (CLADS). The CLADS munition should then be integrated into Volcano (M87A1) canisters to provide a mix of antitank and nonlethal antipersonnel munitions.

Alternatives Potentially Available After 2006

Only well after 2006 will accelerated advancements in technology lead to truly innovative alternatives to APL. As sensor technologies mature into reliable systems of systems, multidimensional sensor networks will become available, which will dramatically improve situational awareness on the battlefield.

Conclusion 8. After 2006, improvements in the tactical effectiveness of existing or proposed remotely delivered antitank (AT) landmines ought to be technologically feasible, which could eliminate the need for mixed systems. Future systems that separate the sensor from the shooter could be improved by multiple means of remote deployment and resistance to countermeasures through signature reduction and other techniques. Track III programs, like the Track I initiative, will require concentrated effort and stable funding. In the long term, the emergence of new technologies, such as the ability to distinguish accurately between combatants and noncombatants, will provide opportunities for the development of systems that can outperform today's antipersonnel landmines.

Alternatives for Use Against Dismounted Targets

The committee considered five systems that should be available after 2006 for use against dismounted targets. When measured against the M14/M16 baseline, they all appeared to meet both the military and humanitarian requirements. All of these systems involved a combination of sensors, communication to a man-in-the-loop, and kill mechanisms. Given their preliminary state of development, the committee did not make any specific recommendations regarding these systems.

Alternatives for Use Against Mounted Targets

The committee considered eight systems that might be available after 2006 for use against mounted enemies. The concepts included enhancing current AT mines by adding nonlethal devices, such as Tasers, to protect them from being breached, or a telemetry and sensor package that could provide near real-time knowledge of the location of scattered minefields or of a breach attempt. The committee also considered the Raptor, a smart, autonomous, AT system already in development that will improve situational awareness and provide targeting information to other weapons, such as the Hornet/WAM.

Recommendation 8a. The Army should proceed rapidly with plans for modernizing existing remotely delivered pure antitank landmine systems, such as the Remote Antiarmor Mine System (RAAMS) and Volcano (M87A1), by incorporating other technologies, including sensors, precision locators, and nonlethal devices.

The Self-Healing Minefield concept, a DARPA Track II program, is an intelligent distributed network of mines with decentralized control. The individual mines detect breaching attempts through mine-to-mine communications and automatically react by moving to fill gaps in the minefield. This innovative system is unlikely to be available in less than 10 years.

Recommendation 8b. The development of the Self-Healing Minefield concept, which automatically reacts to any breaching attempt by refilling gaps, should be experimentally evaluated to determine its operational effectiveness.

The Distributed Web Sensor Complex (DWSC), the focus of a U.S. Army science and technology program, is a sensor network that would exploit future ground-based and air-based combat systems. The concept envisions delivering, by artillery or air, hundreds, or even thousands, of small, expendable sensors over a wide area. Because the DWSC exploits the capabilities of future combat systems and does not require a dedicated kill mechanism, it appears to be one of the most effective future systems, and it scores very high in the military effectiveness category.

Other Considerations

The committee was briefed by representatives of organizations, inside and outside DOD, on concepts and technologies being developed for other purposes, such as non-mine systems, sensors for other defense purposes, and commercial devices. Any of these technologies could be leveraged to provide elements of future alternatives to APL.

Recommendation 8c. Several other technologies or systems already under development for other purposes should be considered as potential components of long-term alternatives to antipersonnel landmines, including unmanned air and ground vehicles, directed-energy weapons, battlefield sensory-illusion devices, passive transponders (e.g., tags), and other lethal and nonlethal systems.

Because U.S. APL, other than some of those used in Korea, are self-destructing and self-deactivating, they do not present as great a danger to noncombatants as do other APL.

Nor do they leave battlefield residue that may inhibit post-war reconstruction. For the safety of both U.S. forces and noncombatants, DOD should consider making other non-recoverable explosive munitions self-destructing and self-deactivating.

Conclusion 9. The self-destructing and self-deactivating capability of today's U.S. scatterable landmines, used in accordance with international law, is a desirable operational capability because it (1) increases maneuver options and (2) addresses humanitarian concerns by reducing residual explosive hazards.

Recommendation 9. Any nonrecoverable, explosive alternative to antipersonnel landmines should have self-destructing and self-deactivating fuzes to meet operational requirements, address humanitarian concerns, and reduce fratricide among friendly troops. The U.S. government should consider equipping all nonrecoverable explosive munitions with similar technologies.

1

Introduction

> The Union Army of the Potomac was pressuring Confederate forces retreating from Yorktown. . . . Suddenly a series of shells exploded beneath the hooves of Federal horses. Pandemonium erupted as many whole Union companies bolted in panic. (Robbins, 1997)

Antipersonnel landmines (APL), an often low-technology, inexpensive staple of armies around the world, became the center of international controversy as the twentieth century drew to a close. On one side of the debate were the military utility of APL and doubts about the feasibility of controlling their use; on the other side were the tragic residual humanitarian effects of APL. In December 1997, 122 countries signed the *Convention on the Prohibition of the Use, Stockpiling, Production and Transfer of Anti-Personnel Mines and on Their Destruction,* known as the Ottawa Convention, banning APL (see Appendix E for text). The Clinton administration announced, "The United States will sign the Ottawa Convention by 2006 if we succeed in identifying and fielding suitable alternatives to our anti-personnel landmines and mixed antitank systems by then" (Berger, 1998). Since then, the United States has destroyed millions of APL that did not have self-destructing or self-deactivating devices. In addition, in accordance with Presidential Decision Directive 64, pure APL (i.e., those that are not part of a mixed APL-antitank [AT] mine system) cannot be used outside Korea after 2003. In the meantime, the United States is searching for alternatives. This National Research Council[1] study was part of the U.S. government's efforts to determine if and when alternatives will be available.

The Committee on Alternative Technologies to Replace Antipersonnel Landmines was created for the purpose of (1) identifying and examining possible tactics, technologies, and operational concepts that could provide tactical advantages similar to those provided by APL by 2006; (2) suggesting a near-term alternative technology, weapon system, or combination of systems that could be derived from known, available systems or that could provide a short-term solution if the recommended alternative will not be available by 2006; and (3) describing how the identified technologies and systems could be used consistently with current tactical doctrine and operational concepts or recommending changes in tactics or operational concepts.

This chapter provides background information on the development and use of landmines, including their use by the U.S. military. It then describes the residual hazards landmines may pose to noncombatants during and after combat and to postwar relief and recovery activities. Various international agreements relating to the use of APL and the evolution of U.S. policy are reviewed as a context for brief descriptions of efforts to identify alternatives. The Statement of Task for this study, a description of the committee process, and a road map for the report are provided at the end of the chapter.

DEFINITIONS

Several accepted definitions for APL are currently in use, leading to confusion over whether a specific APL is compliant with the Ottawa Convention or not. The subtleties and implications of the definitions continue to be the subject of diplomatic and scholarly debate. The U.S. Army doctrinal manual on landmines, *Field Manual 20-32, Mine/Countermine Operations*, uses the following definition:

> A landmine is an explosive device that is designed to destroy or damage equipment or personnel. Equipment targets include ground vehicles, boats, and aircraft. A mine is detonated by the action of its target, the passage of time, or controlled means. There are two types of land-based mines—AT [antitank] and AP [antipersonnel]. (U.S. Army 1998b)

Definitions in various treaty documents that specifically address APL are shown in Box 1-1.

For the purposes of this study, the committee used the definition found in the *Convention on Conventional Weapons (CCW), Amended Protocol II*, an international convention that has been signed and ratified by the United States

[1] The National Research Council is the operating arm of the National Academy of Sciences, the National Academy of Engineering, and the Institute of Medicine. Together these institutions are known as the National Academies.

BOX 1-1

Definitions of Antipersonnel Landmines

Convention on Conventional Weapons (CCW), Amended Protocol II

Antipersonnel mine means a mine primarily designed to be exploded by the presence, proximity, or contact of a person and that will incapacitate, injure, or kill one or more persons

Ottawa Convention

Antipersonnel mine means a mine designed to be exploded by the presence, proximity, or contact of a person and that will incapacitate, injure, or kill one or more persons. Mines designed to be detonated by the presence, proximity, or contact of a vehicle, as opposed to a person, that are equipped with antihandling devices are not considered antipersonnel mines as a result of being so equipped.

(see Appendix E for text). APL that are part of mixed systems fall under this definition; antihandling devices[2] do not.

HISTORY OF MINES

The use of mine-like devices has a long history in military operations,[3] but widespread concerns have arisen only recently, primarily because of the increasing proliferation of mines. A tenet of military operations is to force the enemy into a disadvantageous position by controlling terrain and the situation on the battlefield while conserving combat power. Ideally, all weapons used in war are designed to provide friendly forces with maximum flexibility and to inflict maximum damage on the enemy. In recent years, a concerted effort has been made to reduce the effects of all weapons on noncombatants—so-called collateral damage.

The first landmines in the West (in the sixteenth century) required high maintenance and were susceptible to dampness. By the nineteenth century, the availability of explosive shells and the invention of the percussion cap enabled the development of more water-resistant mines (Schneck, 1998).

APL were first used in the American Civil War by the Confederate Army during the Peninsula Campaign of 1862. Developed by Brigadier General Gabriel Rains and known as land "torpedoes" or as the subterra explosive shell, these APL would explode when a soldier (or a horse or wagon) stepped on the fuze. Although they were decried by General McClellan of the Union Army, similar devices were used by General Sherman during his March to the Sea. The idea of marking mines with small flags planted 10 feet in front of them on the defender's side was introduced at this time. Explosive mining tunnels under fortified positions were used at Vicksburg in 1863 and again at Petersburg in 1864. U.S. armies did not use mines again for 76 years (Croll, 1998).

Landmines were used between 1865 and 1914 by Prussia (1870), the British (in numerous colony wars), and Russia (1904). In response to the introduction of tanks by the British in World War I, the Germans fabricated explosive AT mines, improvised in the field from artillery projectiles. Later, mines were mass-produced to improve their efficiency. By the end of the war, both sides had a small inventory of AT mines (Croll, 1998). As early as 1918, the Germans had developed a methodology for laying minefields in a pattern, marking and recording them, and protecting them with observation and small-arms fire. Soon thereafter, the Allies also initiated a doctrinal requirement that minefields be marked and recorded.

In World War II, landmines were widely used as a counterforce to the inherent mobility of large armored formations. Concurrently, smaller APL were developed to discourage foot soldiers from disabling the AT mines[4] and for use in terrain where infantry forces predominated.

The Germans, who developed extensive mine warfare practices based on their antitank operations in World War I, had refined their methods for laying mines during the interwar period. Mines were typically laid in a uniform pattern; the friendly side of a minefield was usually marked, as were lanes and cleared areas; and locations of minefields were recorded. The minefields were observed and protected with covering fire from antiarmor weapons, small arms, and artillery.

Although no new aspects of mine warfare were introduced in the Korean War, the lessons learned in World War II were tested and affirmed. Mines were used during the Korean Conflict to cover withdrawals and to reinforce defenses. However, United Nations forces did not always mark and record minefields, which sometimes resulted in casualties to friendly forces crossing unmarked minefields. In some

[2] Antihandling devices perform the function of a mine fuze if someone attempts to tamper with the mine. They are intended to prevent someone from moving or removing the individual mine, not to prevent reduction of the minefield by enemy dismounts. An antihandling device usually consists of an explosive charge that is connected to, placed next to, or manufactured in the mine. The device can be attached to the mine body and activated by a wire that is attached to a firing mechanism. U.S. forces can use antihandling devices only on conventional AT mines (U.S. Army, 1998b).

[3] As early as the ninth century B.C., the Assyrian army dug tunnels under walls and fortifications, creating breaches when the wooden beams supporting the tunnels were set on fire and the ground above them collapsed. The development of gunpowder by the Chinese in the ninth century and its later production and use in Europe led to more effective mining (Schneck, 1998). The term "*mines*" in reference to an explosive charge in or on the ground is derived from these tunneling (or mining) activities.

[4] This action has come to be called a dismounted breach.

cases, unguarded minefields were removed and reused by the enemy against United Nations forces (Roy and Friesen, 1999). Allied forces found that they could not always stop waves of attackers willing to take the significant casualties caused by forcing their way through an active minefield. After the war, the U.S. military called for a light APL that would guarantee casualties, which led to the development of the M14 pressure mine and the M18 (Claymore) (Croll, 1998). United Nations forces were confident of the efficacy of using APL to supplement other, more lethal means of defense (Roy and Friesen, 1999) and South Korean and American forces laid extensive minefields on the border between North and South Korea to deter or delay an invasion. The growing importance of landmines was also evidenced by the emphasis on them in military field manuals written after the Korean War. Many also believed that mines could help stop invading Warsaw Pact armies if a war broke out in Europe.

The developments in mine warfare after World War II and Korea, especially the use of protective minefields to guard well defined areas, such as borders, were of little use to the United States in Vietnam. Vietnamese, U.S., and allied forces operated from base camps or fortified enclaves throughout the country, but the insurgents, and later the North Vietnamese, moved throughout the countryside. Depending on the enemy's objectives, the terrain, and other factors, he was capable of conducting both standoff attacks and ambushes.

When U.S. combat units arrived and established base camps, minefields were emplaced in many locations to provide security perimeters. These minefields were a combination of conventional nonself-destructing APL (i.e., M14s and M16s), trip flares, and Claymores (M18s). In some cases, improvised mines were also used. Within a short time, because of the buildup of forces, many base camp perimeters were expanded, which necessitated the clearance or isolation of previously emplaced APL. U.S. forces also aggressively patrolled outside their base camps, which posed the problem of crossing defensive minefields. Therefore, early in the war, the use of APL in large defensive minefields around U.S. base camps was drastically reduced. Later, during the Vietnamization of the war, when the United States had become less active offensively, mines were again used to protect bases and camps.

The APL most used by U.S. and allied forces in Vietnam was the M18 series Claymore, used extensively around base camps and to protect positions established in the field, as well as on ambushes by all combat units. The Claymore was a basic component of every infantryman's gear.

Although the United States developed several experimental mines specifically for use in Vietnam, none was permanently adopted. For example, the XM-61, a linear explosive charge (similar to detonating cord) wrapped at intervals with a fragmentation sleeve, was developed for use as a command-detonated mine along trails during ambushes. Several air-delivered mines were also introduced, including the "Gravel Mine" (XM42 mine dispensing system), the BLU-42/B APL, BLU-43/B APL, and the BLU-45 (the first scatterable AT mine). The BLU-43/B, also called the "Dragontooth," was filled with liquid explosive and detonated by the application of about 7 kilograms of weight. Although the BLU 43/B was never adopted as a standard service item by the United States, it was copied by the Soviets (PFM-1 and PFM-1S, called the Butterfly) and used in large quantities in Afghanistan.

A major challenge for U.S. forces in Vietnam was countermine activities to minimize the use of mines by enemy forces. Mines became a constant threat during U.S. convoy operations to resupply base camps. Command-detonated mines, either locally fabricated or made from unexploded ordnance, such as artillery projectiles and aerial bombs, were buried beneath and alongside roads. Finding them and removing them was a daily challenge.

The Viet Cong made extensive, effective use of mines and booby traps to protect their base areas and target paths and roads. Like the Irish Republican Army later in Northern Ireland, the Viet Cong used command-detonated and timer-detonated mines in populated areas as terrorist weapons against military and civilian targets. North Vietnamese Army units used mines in generally the same way as their Viet Cong counterparts. By most accounts, this use of mines and booby traps inflicted a much higher percentage of casualties in Vietnam than it did in World War II or Korea and had even more significant psychological effects.[5]

In the limited wars of the 1970s and 1980s, landmines continued to be used, sometimes effectively and sometimes not. Guerilla forces in Mozambique, Angola, and Rhodesia showed that mines could be used effectively for "unconventional" warfare, to instill terror in the population or to force migrations by making an area uninhabitable. The speed of operations during the Yom Kippur War (1973) demonstrated that traditional minefields or mine tactics could sometimes be a hindrance to one's own movements. Subsequently, many countries began working to improve their mines for use in rapidly paced operations.

In the early 1980s, the first self-neutralizing systems with a selection of self-destruct times were deployed. The Italians developed helicopter-delivered, scatterable mine systems. Although these were an improvement over manually placed mines, the helicopters were susceptible to ground fire. From these early systems, the United States developed the "family of scatterable mines" (FASCAM), which can be delivered by ground launcher, helicopter, fixed-wing aircraft,

[5] According to sources cited by Roy and Friesen (1999), wounds and deaths in Vietnam from mines and booby traps were 11 percent and 15 percent, respectively, of all casualties in Vietnam, compared with 3 to 4 percent in World War II and Korea. The authors also quote one account of small unit actions in Vietnam: "The enemy they found hardest to combat was not the VC; it was mines."

or artillery. FASCAM devices that can be deployed rapidly and have self-destructing/self-deactivating mechanisms have been integrated into current U.S. doctrine for fast-paced, maneuver warfare.

The FASCAM were first deployed during the Gulf War ahead of and behind Iraqi positions to prevent the movement of forces. They were also used against airfields and storage facilities for chemical weapons. The Iraqis, drawing in part on their experience in the war with Iran in the 1980s, began laying extensive, traditional minefields of both AT mines and APL immediately after their occupation of Kuwait. The coalition forces, in turn, made extensive preparations to deal with the minefields by eliminating Iraqi overwatching fires and by thoroughly preparing to conduct countermine operations. The offensive campaign plan was built around countering the Iraqi minefields and effectively neutralized their impact on coalition forces. As a result, the Iraqis were unable to stop, or even appreciably slow, the ground attack against them. Success was attributed to many factors, including the coalition's ability to survey minefields and exploit their weaknesses and the Iraqis' inability to observe and defend their barriers.

RESIDUAL HAZARDS OF MINES

Nonself-destructing landmines, even when used according to the generally accepted Western doctrine of marking and recording minefields, can continue to pose hazards after conflicts have ended. The danger of nonself-destructing mines is part of the more general problem of unexploded ordnance, although hazards from landmines are particularly serious because they are target activated. In 1960, five Civil War era landmines found in Alabama were determined to be capable of exploding on contact. Since the 1940s, several active mines from World War II have been found each year scattered throughout Western Europe. After the Gulf War, landmines posed a hazard for both soldiers and civilians. Unfortunately, until the 1980s, most mines were not self-deactivating or self-destructing.

The adoption of self-destructing and self-deactivating devices by the United States, which, when used in accordance with accepted international practice, can largely eliminate residual hazards, was copied only by its NATO allies and a few other countries.[6] Therefore, the bulk of the mines in use around the world are still nonself-deactivating or nonself-destructing APL. Because they are cheap and easy to obtain, they are especially attractive to regional and nonstate belligerents.

It is impossible to estimate accurately the number of landmines around the world today. The Office of Humanitarian Demining Program of the U.S. Department of State has used the figure of 60 to 70 million landmines in more than 60 countries, mostly in Africa, the Middle East, Southeast Asia, South America, and Latin America. The International Committee of the Red Cross estimates that during the late 1980s and early 1990s landmines killed or maimed more than 25,000 people each year (Patierno, 2000a). However, this figure may include a large number of casualties caused by unexploded ordnance.

In addition to causing casualties, the presence of landmines can also seriously inhibit relief efforts during conflicts and resettlement and reconstruction after conflicts are over. Mine-infested land creates refugees and prevents the resettlement of people who fled during the conflict. Even a suspicion that fields are mined may render them unusable. If a country's infrastructure (roads, bridges, and railroads) has been mined, economies are much more difficult to rebuild. The burden of uncleared landmines on war-devastated countries was a major motivation for trying to prevent the emplacement of new mines. For many international organizations, governments, and nongovernmental organizations, the humanitarian costs of mines outweigh their military advantages.

INTERNATIONAL INSTRUMENTS

For centuries the international community has attempted to minimize unnecessary wartime suffering by combatants and noncombatants alike. Beginning in the mid-1800s, the increasing destructiveness of weapons made possible by industrialization made the problem more urgent. The 1864 *Geneva Convention for the Amelioration of the Condition of the Wounded in Armies in the Field* introduced principles upheld in later Geneva Conventions that the wounded be treated humanely regardless of nationality and that medical personnel and units be regarded as neutral. In 1899, and again in 1907, the Hague Peace Conferences reaffirmed the laws and customs governing land warfare. As warfare was extended to the air, concerns about the treatment of civilians in enemy territory increased, and in 1949 the international community addressed the issue of the safety of noncombatants in the *Geneva Convention (IV) Relative to the Protection of Civilian Persons in Time of War* (ICRC, 1949).

Throughout this report, the committee uses the term *humanitarian,* which can be construed to have more than one meaning. The principal meaning is the effort to protect noncombatants from the effects of wartime weapons. As, a corollary, humanitarian also refers to compliance with the international agreements limiting or banning mines. The humanitarian intent of these instruments was the basis for the committee's development of the humanitarian criteria for alternatives to APL described in Chapter 4.

Convention on Conventional Weapons

From time to time, the international community has attempted to regulate the possession or use of a weapon or

[6] This has been attributed to their greater complexity and higher production and acquisition costs.

even to ban it completely. The 1996 *Chemical Weapons Convention* and the *1972 Biological and Toxin Weapons Convention* are recent examples of the prohibition of a weapon. The 1980 *Convention on Prohibitions or Restrictions on the Use of Certain Conventional Weapons Which May Be Deemed to Be Excessively Injurious or to Have Indiscriminate Effects* (CCW) was the first international treaty to attempt to regulate the use of landmines specifically. The *Protocol on Prohibitions or Restrictions on the Use of Mines, Booby-traps and Other Devices (Protocol II)* proscribed the use of APL against civilians or in areas of civilian settlement, as well as their indiscriminate use (i.e., not directed at a military objective). The CCW required accurate recording of all mines to facilitate their removal, and prohibited the use of remotely delivered mines unless they had neutralizing mechanisms or their locations could be accurately recorded. The protocol also called for agreements after the cessation of hostilities, among the parties to the conflict and with other states and international organizations, as necessary, to remove all mines or render them ineffective. Shortcomings of the protocol included: (1) it was not applicable to internal conflicts; (2) it did not provide a probation period for modifying non-detectable APL, and (3) it did not cover long-lived APL (Matheson, 1999).

After an extensive review of the CCW, an amended landmine protocol was issued in May 1996 addressing these shortcomings (see Appendix E for the text). Amended Protocol II, which entered into force on December 3, 1998, distinguishes between APL and AT mines and further restricts the use of mines and minefields. All APL must be detectable (i.e., manufactured or modified with a minimum amount of metal content specified in the Technical Annex to the Protocol). All remotely emplaced APL must be equipped with self-destructing devices and backup self-deactivating devices. All nonremotely emplaced APL must either be placed in an area with a marked and monitored perimeter or must be equipped with self-destructing and self-deactivating devices. Amended Protocol II also established rules governing the transfer of landmines and extended the protocol to cover the use of landmines in internal conflicts. As of June 15, 2000, 50 countries were party to the Amended Protocol II (79 were party to the CCW).

The CCW, including Amended Protocol II, will be reviewed again in 2001. The United States envisions further improvements to the protocol with respect to applying the APL detectability standards to AT mines, the adoption of increased self-destruction and self-deactivation requirements, and adding a verification and compliance mechanism (Matheson, 1999).

The Ottawa Convention

At the same time the CCW was undergoing review, nongovernmental organizations began to address the issue of the use and humanitarian consequences of landmines. In 1992, six nongovernmental organizations (Handicap International, Human Rights Watch, Medico International, Mines Advisory Group, Physicians for Human Rights, and Vietnam Veterans of America Foundation) joined forces to create the International Campaign to Ban Landmines, which called for a total ban on the use, production, stockpiling, and transfer of APL (ICBL, 2000). National campaigns in several countries followed, many nongovernmental organizations worldwide[7] joined in the fight, and the International Campaign to Ban Landmines gradually gained the support of several like-minded governments, most notably Canada.

In October 1996, the Canadian government hosted the *Towards a Global Ban on Landmines: International Strategy Conference*, which was attended by representatives of 74 countries. At the conclusion of the meeting, 50 government participants agreed to a statement expressing the need for a ban on APL and Canada announced it would hold a treaty-signing conference for a total ban in December 1997. Preparatory conferences to discuss and develop the text of a draft treaty, prepared initially by Austria, were held in Vienna in February 1997, Bonn in April 1997, and Brussels in June 1997. The treaty was negotiated over a three-week period in September 1997 in Oslo, Norway. On December 3, 1997, 122 nations signed the *Convention on the Prohibition of the Use, Stockpiling, Production and Transfer of Anti-Personnel Mines and on Their Destruction* (called the Ottawa Convention or Mine Ban Treaty) in Ottawa (see Appendix E).

The Ottawa Convention bans the use of APL under any circumstances. The ban includes APL used alone, APL used in mixed systems, and APL that are self-destructing and self-deactivating.[8] Furthermore, it prohibits the development, production, or any other means of acquisition, stockpiling, retention, or transfer of APL to anyone, directly or indirectly. Governments that sign agree not to assist, encourage, or induce, in any way, anyone to engage in any activity prohibited to a state party under this Convention. Finally, each signatory must undertake to destroy or ensure the destruction of all APL in accordance with the provisions of the Convention.

By September 1998, 40 countries had ratified the convention, thus bringing it into force as international law on March 1, 1999. As of September 7, 2000, a total of 139 nations had signed or acceded to the Ottawa Convention, including all NATO member states, except the United States and Turkey, and all European Union member states, except Finland. Of the 139, 107 have ratified the convention (see Appendix F).

[7] Currently, there are over 1,100 organizations in over 60 countries that are part of the campaign network (ICBL, 2000).

[8] The negotiators did not allow for the inclusion of self-destructing and self-deactivating APL for several reasons. These mines still fit the definition of an APL and no exceptions were to be made. If an exception had been made for these mines, primarily in the inventory of only the United States and a few western European countries, exceptions might have had to be made for weapon systems of other countries.

A number of major mine producers or nations in regions of conflict, including Russia, China, Egypt, Israel, India, Pakistan, and North and South Korea, have not agreed to the treaty.

THE U.S. POSITION

Although the use of landmines by U.S. forces did not create the current humanitarian crisis, the U.S. government has taken strong actions toward mitigating the effects of indiscriminate use of APL around the world. These actions include a ban on exports, assistance with clearance of mines (also called demining), assistance to victims, and a search for alternatives to APL.

Ban on Exports

As a result of legislation introduced by Senator Patrick Leahy (D-Vermont), the United States has had a moratorium on exports of APL since 1992 (Rieser, 1999). The first one-year moratorium on exports became law on October 23, 1992, as part of the National Defense Authorization Act for Fiscal Year 1993. The United States thus became the first country to enact legislation controlling APL. The moratorium was extended to four years in the National Defense Authorization Act for Fiscal Year 1994 and to five years in the Foreign Operations, Export Financing, and Related Programs Appropriations Act for Fiscal Year 1996. On January 17, 1997, President Clinton announced that the United States would permanently ban the export and transfer of APL (White House, 1997a).

Mine Clearance and Assistance to Victims

Since 1988, the United States has been assisting countries affected by landmines in several ways: promoting awareness (educating people about the dangers of landmines and what to do when they are found); actively searching for mines and clearing minefields; and providing assistance to victims. Since 1993, the United States has provided assistance to more than 35 countries (Patierno, 2000b).

The Humanitarian Demining Program was created in 1993 to establish self-sustaining, indigenous demining programs, reduce civilian casualties, facilitate the return of refugees, enhance the stability of affected nations, and encourage international cooperation and participation. The Interagency Working Group on Humanitarian Demining, chaired by the U.S. Department of State with the U.S. Department of Defense (DOD) as vice-chair, is responsible for approving, developing, and coordinating U.S. humanitarian demining policies and programs (DOS, 2000). Since 1994, the U.S. military has trained indigenous forces in demining techniques and assisted in the establishment of in-country training programs.

Movement Toward a Ban

President Clinton first called for the elimination of APL in a speech to the United Nations (UN) General Assembly on September 26, 1994 (White House, 1994). On May 16, 1996, he announced a new policy, including a commitment to pursue an international ban on APL and to destroy about three million nonself-destructing APL by the end of 1999, retaining only those necessary for training and for defense of the Demilitarized Zone in Korea (White House, 1996). On December 10, 1996, the UN General Assembly voted (156-0) in favor of a U.S.-initiated resolution urging states to pursue an agreement to ban APL.

In the meantime, the United States continued to work toward limiting the use of landmines. On January 7, 1997, the president transmitted the CCW Amended Protocol II to Congress for ratification; Congress ratified it on May 24, 1999. The United States had also planned to work toward a worldwide treaty banning APL through the United Nations Conference on Disarmament, which largely deals with nuclear matters and operates on a consensus rule (White House, 1997a). This initiative did not elicit support from other members of the conference.

The Ottawa Convention

Despite America's strict limitations on APL and its initial support for a ban, the United States has not signed the Ottawa Convention. The U.S. government had expressed both general and specific concerns throughout the negotiations, but in the end, two primary concerns dominated. As President Clinton explained on September 17, 1997, for the United States to sign the treaty, two provisions would have to be included. First, the United States wanted a transition period during which APL could be phased out to ensure that enough time would be available to devise alternatives. Second, the United States wanted to preserve its mixed AT mine systems, which include APL, as additional protection against dismounted breaching (Clinton, 1997; Witkowsky, 1999).

The decision not to sign the Ottawa Convention was strongly influenced by security concerns on the Korean Peninsula (Witkowsky, 1999). The U.S. government and the U.S. military were convinced that APL, including APL without self-deactivation/self-destruction mechanisms and APL in mixed systems, are essential to the defense of the Demilitarized Zone in Korea. "The security situation in Korea is unique, requiring the United States to maintain the option of using [APL] there until alternatives are available or the risk of aggression has been removed" (White House, 1997b). As General John H. Tilelli, then Commander in Chief (CINC), UN Command/Combined Forces Command (UNC/CFC), and Commander, U.S. Forces Korea, testified before the House Armed Services Committee:

> ... these weapons, both the non-self-destructing and self-deactivating types, are absolutely vital to the success of UNC/CFC's

> mission to deter North Korea aggression and defend the [Republic of Korea]. (Tilelli, 1999)

Although the United States would not sign the Ottawa Convention in 1997, the president announced further refinements to U.S. policy as a demonstration of the U.S. commitment to ending the use of all APL:

> ... I'm directing the Department of Defense to develop alternatives to antipersonnel land mines so that by the year 2003 we can end even the use of self-destruct land mines . . . everywhere but Korea. As for Korea, my directive calls for alternatives to be ready by 2006, the time period for which we were negotiating in Oslo. By setting these deadlines, we will speed the development of new technologies.... In short, this program will eliminate all antipersonnel land mines from America's arsenal. (Clinton, 1997)

President Clinton had repeatedly stated that the U.S. goal is to sign the Ottawa Convention as soon as its concerns have been met. Assistant to the President for National Security Affairs Samuel Berger reiterated this commitment in a letter to Senator Leahy stating that the United States would search for alternatives for Korea and for mixed AT mine systems, including replacements for APL in mixed systems or replacements for the mixed systems entirely. If both alternatives (for Korea and for mixed systems) could be found by 2006, the United States would sign the Ottawa Convention (Berger, 1998). If the United States decides to sign the Ottawa Convention, all of the APL in the current inventory, except the command-detonated Claymore, would be banned and would have to be destroyed within four years of the signing.

Search for Alternatives

In a memo dated October 21, 1997, the deputy secretary of defense assigned the Office of the Under Secretary of Defense (Acquisition, Technology and Logistics) to develop and oversee a two-track program to find alternatives to APL. Track I, headed by the secretary of the Army, has two major components: (1) the search for an alternative to the nonself-destructing APL designated for use in Korea and (2) the development of Remote Area-Denial Artillery Munition (RADAM) for use in Korea until 2006 (or when suitable alternatives are available).[9]

The purpose of Track II, headed by the Defense Advanced Research Projects Agency (DARPA), is to develop "alternatives to meet the requirements currently met by APL" and to "investigate maneuver denial approaches that may be more innovative and/or take advantage of advanced technologies" (Altshuler, 1999). DARPA's current efforts are focused on the development of a self-healing minefield and tags/minimally guided munitions (for a description and assessment of each of these, see Chapter 7).

In June 1998, Presidential Decision Directive (PDD) 64 added the requirement that DOD "actively investigate alternatives to the anti-personnel submunitions used in our mixed anti-tank systems, as well as actively explore possible replacements for all mixed munitions" (DOD, 2000). The candidate alternatives must be militarily advantageous, cost effective, and safe. PDD 64 did not include a schedule.

A memorandum on March 23, 1999, from the deputy secretary of defense created the Track III program under the leadership of the Office of the Under Secretary of Defense (Acquisition, Technology and Logistics). The initial purpose of Track III was to address "doctrine, tactics, force structure options, use of combat systems currently fielded or under development, Track 1 and 2 alternatives, materiel and non-materiel alternatives, and alternatives recommended by the combatant commanders." The first phase of concept evaluation was to assess alternatives that could be developed and fielded for the near term (by 2006), the midterm (2006 to 2012), and the long term (beyond 2012) (DOD, 2000).

Track III guidance was eventually focused on the development of materiel or nonmateriel alternatives to replace all landmines. Nicknamed RATTLER (rapid tactical terrain limiter), a number of working groups were formed to carry out the first phase of the Track III program. An array of 74 refined ideas were gradually reduced to 22 idea categories, then 17 initial concepts, nine combined concepts, and eventually seven final concepts, three of which are undergoing concept exploration by industry (and are considered proprietary). RATTLER projects are exploring three principal concepts: (1) the use of sensors that are not co-located with the effects (lethal or nonlethal responses) that require "just-in-time" delivery or prior placement of the responding weapons; (2) sensors and effects (lethal and nonlethal) that are co-located; and (3) sensors that are used for situational awareness and are protected by AT mines (Morelli, 2000).

The identification of alternatives for landmines will require considerable effort. To date, progress on Track I has been slow and technologies being explored under Track II will not be available until well after 2006. The concepts developed under the first phase of Track III are too new for the committee to judge their potential.

Recommendation. If the decision is made to accede to the Ottawa Convention, a transition period may be necessary before implementation to maintain current U.S. military capabilities until suitable alternatives can be made available. During that transition, existing self-destructing and self-deactivating antipersonnel landmines should be retained, both in their stand-alone form and as part of mixed systems.

Recommendation. Of the solutions not compliant with the Ottawa Convention, simply retaining the current

---

[9] The Remote Area-Denial Artillery Munition (RADAM) would combine AT mines (the Remote Antiarmor Mine System [RAAMS]) and APL (the Area-Denial Artillery Munition [ADAM]) into a single projectile to create a new mixed system. Although RADAM fulfills the requirement set by President Clinton to eliminate pure APL everywhere but Korea by 2003, it would not be compliant with the Ottawa Convention because it contains APL (see Chapters 5 and 6).

self-destructing and self-deactivating mines would be the best course of action.

Funding for Alternatives

The progress report by DOD to Congress on the search for alternatives included a table (Table 1-1) showing current and projected funding levels (DOD, 2000).

Congressional Mandate

In addition to providing direction and funding for the search for alternatives, the Strom Thurmond National Defense Authorization Act for FY99 (Public Law 105-261, Section 248) and the conference report accompanying the 1999 Department of Defense Appropriations Act (H.R. 105-746) mandated that the secretary of defense enter into contracts with two scientific organizations to study alternatives to APL. DOD concluded contracts with (1) a U.S. Department of Energy team of Lawrence Livermore and Los Alamos National Laboratories and (2) the National Academy of Sciences. These studies are part of the Track III effort. The following statement of task was agreed to between DOD and the National Academy of Sciences for work to be carried out by its operating arm, the National Research Council.

> The National Research Council, building on its experience in examining and assessing potential technologies for military application, will appoint a committee that will:
>
> 1) Identify and examine possible alternative tactics, technologies, and operational concepts for APL capable of providing similar tactical advantages for U.S. forces. A solution should be selected for its ability to meet the future warfighting needs of U.S. forces and provide the strategic and tactical benefits of the system it is replacing. If selected alternatives require research and development to the extent that they will not be available to implement before 2006, the committee will suggest a near-term alternative technology, weapon system, or combination of systems that can be derived from known, already available systems or systems of systems that can act as a near term solution until the long-term solution is available.
>
> 2) Describe how the identified technologies/systems will best be applied for U.S. force use consistent with current tactical doctrine and operational concepts, or what changes in tactics or operational concepts would be required to achieve comparable results.

COMMITTEE PROCESS

The use of APL is a sensitive and contentious political and military issue. Therefore, in creating the Committee on Alternative Technologies to Replace Antipersonnel Landmines, the National Research Council (the operating arm of the National Academy of Sciences) selected committee members representing a broad spectrum of backgrounds, expertise, and interests. Areas of expertise include technology development, experimental design, military operations, and defense policy (see Appendix A for biographies).

In addition, the committee relied on the expertise and advice of representatives of the National Security Council, the U.S. Department of State, DOD, industry, and several nongovernmental organizations. Classified information was provided on several occasions to ensure familiarity with as many aspects of the search for alternatives as possible. Information available in the open literature and material submitted by experts, as well as the practices of nations that have already signed the Ottawa Convention (see Appendix F), were reviewed. Meetings of the full committee were complemented by site visits by smaller subcommittees (Appendix B).

Even though the Statement of Task did not require the committee to consider the Ottawa Convention, the committee concluded that this study would not have been empanelled were it not for the Ottawa Convention and attendant humanitarian concerns. The committee believed the major reasons for seeking alternatives to APL are humanitarian concerns, compliance with the Ottawa Convention, and enhanced military effectiveness. The current inventory of

TABLE 1-1 Current and Projected Funding for Tracks I, II, and III (in $ millions)

Project	FY98	FY99	FY00	FY01	FY02	FY03	FY04	FY05	TOT
Track I	2.7	17.2	18.2	12.5	67.2	121.8	121.6	121.4	482.6
Track II DARPA			7	9.9					16.9
NSD-A Studies		2							2
Track III	2		19.6	26.4	29.1	29.2	48.7	77.8	232.8
TOTALS	4.7	19.2	44.8	48.8	96.3	151.0	170.3	199.2	734.3

SOURCE: DOD, 2000.

NOTE: Figures are from the president's FY01 Budget Submission.

self-destructing and self-deactivating U.S. APL is militarily advantageous and safe (i.e. they achieve desired military objectives without endangering U.S. warfighters or noncombatants more than other weapons of war), but they are not compliant with the Ottawa Convention. However, humanitarian concerns and Ottawa compliance are not always synonymous. In fact, some of the apparently Ottawa-compliant alternatives examined by the committee may be less humane than present U.S. self-destructing and self-deactivating landmines. Therefore, the committee spent a considerable amount of time clarifying and defining the framework of the treaty and included complying with the Ottawa Convention as a criterion for evaluating alternatives.

REPORT ROAD MAP

Chapter 2 characterizes the current and future national security environments and describes how the functions served by landmines might change with technological advances in weaponry. Chapter 3 describes the current uses of landmines. Chapter 4 explains the committee's selection criteria and methodology for analyzing alternatives. Chapter 5 analyzes currently available technologies that might provide the same capabilities as APL. Chapter 6 evaluates technologies and alternatives that should be available by 2006. Chapter 7 analyzes technologies and proposes alternatives that might be available after 2006. Chapter 8 is a complete list of conclusions and recommendations. Appendix A contains the biographies of committee members. Appendix B lists meetings of the full committee, site visits, and meetings of subcommittees. Appendix C gives a description of current types of landmines. Appendix D provides information about how minefields can be breached and discusses the value of APL in protecting minefields from breaching. The texts of the CCW Amended Protocol II and the Ottawa Convention are reprinted in Appendix E. Appendix F lists the countries that have signed and/or ratified the Ottawa Convention and any work on alternatives to APL by other countries of which the committee was aware. Appendix G provides copies of DOD's mission need statements for APL alternatives.

2

National Security Environments and the Context for Landmines

> ...among the most deplorable developments...are (1) extensive use of antipersonnel mines in the conflicts in Chechnya and Kosovo, especially by Russian and Yugoslav forces, and (2) continued use of antipersonnel mines by treaty signatory Angola, and likely use of antipersonnel mines by treaty signatories Burundi and Sudan. (Human Rights Watch, 2000)

Current national security and potential warfighting environments are complex and are expected to become even more so. Although the United States is nominally at peace, large numbers of U.S. forces are currently deployed around the world, either as deterrents to aggression or as peacekeepers. This chapter describes the international environment through the lens of planning documents used to prepare for an event or series of events in which U.S. military power will have to be used. Even though landmines, which are essentially tactical weapons, are seldom mentioned, a description of these documents will provide the reader with an idea of situations in which mines would be used. The chapter begins with descriptions of current strategies, such as the national security strategy, the defense strategy, and efforts by various military services to implement these strategies. This is followed by the views of some of the U.S. commanders responsible for protecting U.S. interests abroad. Also described are the technological opportunities provided by the ongoing revolution in military affairs. The chapter concludes with a brief description of the benefits and vulnerabilities of several advanced technology concepts.

NATIONAL SECURITY STRATEGIES

The strategies described below provide a context for how APL may be used in the future. The committee reviewed many strategy documents and was briefed by several military leaders on how future operations might be conducted. In addition, the committee met with representatives of the National Security Council and the U.S. Department of State.

A National Security Strategy for a New Century

Every year, as required by law, the president issues a national security strategy defining the United States' vital national interests and how they should be protected. In December 1999, President Clinton issued his annual update, *A National Security Strategy for a New Century,* which describes anticipated threats to the United States. The strategy's core objectives are to enhance America's security, bolster America's economic prosperity, and promote democracy and human rights abroad. The president stated that "arms control and nonproliferation initiatives are an essential element of our national security strategy," and the strategy referred specifically to APL:

> [The United States is] committed to ending the threat to innocent civilians from antipersonnel landmines (APL). The United States has already taken major steps toward this goal while ensuring our ability to meet international obligations and provide for the safety and security of our men and women in uniform. President Clinton has directed the Defense Department to end the use of all APL, including self-deactivating APL, outside Korea by 2003 and to pursue aggressively the objective of having APL alternatives ready for Korea by 2006. We will also aggressively pursue alternatives to our mixed anti-tank systems that contain antipersonnel submunitions. We have made clear that the United States will sign the Ottawa Convention by 2006 if by then we have succeeded in identifying and fielding suitable alternatives to our self-deactivating APL and mixed anti-tank systems. (White House, 1999)

The national security strategy envisions that the U.S. military will be faced with an array of threats to our interests, including direct threats to the continental United States, small-scale contingencies, major theater wars, terrorism, cyber attack, information operations, and the threat or use of weapons of mass destruction.

U.S. Department of Defense Strategy

In his 1999 *Annual Report to the President and the Congress,* the secretary of defense affirmed the value of a strong military, stating that the essence of the U.S. defense strategy between now and 2015 will be shaping the international security environment, responding to the full spectrum of crises, as required, and preparing now for an uncertain future (DOD, 1999).

Echoing presidential concerns about current instability and anticipating the advantages that will accrue from U.S. scientific and technological superiority, the secretary of defense stated:

> The fundamental challenge confronting the Department of Defense is simple but daunting. U.S. armed forces must meet the immediate

> demands of a dangerous world by shaping and responding throughout the next 15 years, while at the same time transforming U.S. combat capabilities and support structures to be able to shape and respond effectively in the face of challenges in the future. (DOD, 1999)

To carry out its strategy, the DOD will take the following actions:

- Pursue a focused modernization program to replace aging systems and incorporate cutting-edge technologies to ensure continued U.S. military superiority.
- Continue to exploit the revolution in military affairs to improve the U.S. military's ability to perform near-term missions and meet future challenges.
- Exploit the revolution in business affairs to reengineer DOD's infrastructure and support activities.
- Ensure against unlikely, but significant, future threats so that risk in a resource-constrained environment can be managed effectively, and position the military to respond quickly and effectively to new threats as they emerge.

Joint Vision 2010 and *Joint Vision 2020*

Joint Vision 2010, issued by the Chairman of the Joint Chiefs of Staff (CJCS) in 1996, is consistent with both the president's national security strategy and DOD's strategy. *Joint Vision 2010* provides:

> ...the conceptual template for how America's armed forces will channel the vitality and innovations of our people and leverage our technological opportunities to achieve new levels of effectiveness in joint warfighting. (CJCS, 1996)

Based on assumptions about emerging information-age technologies, the strategy in *Joint Vision 2010* depends on unprecedented cooperation between the Army, Marine Corps, Navy, and Air Force (Close, 1999).

Joint Vision 2020, issued in June 2000, builds on and extends the conceptual template established by *Joint Vision 2010* to guide the continuing transformation of U.S. forces. The goal of the transformation is the creation of a force that is dominant across the full spectrum of military operations, based on the strategic concepts of decisive power, power projection, overseas presence, and strategic agility. Although considerable emphasis is placed on information operations, "...information superiority neither equates to perfect information, nor does it mean the elimination of the fog of war" (CJCS, 2000). *Joint Vision 2020* is based on the following assumptions:

- The United States will continue to have global interests and to be engaged with a variety of regional actors.
- Potential adversaries will have access to the global commercial industrial base and much of the same technology as the U.S. military.
- Potential adversaries will be able to adapt as U.S. capabilities evolve.

Comments of Regional Commanders-in-Chief

Strong statements in support of the continued use of landmines were made in the testimonies of two regional CINCs before the U.S. Congress. General John H. Tilelli, Jr., then CINC of the United Nations Command, Combined Forces Command, and U.S. Forces Korea, testified that that "these weapons...are absolutely vital to the success" of the U.S. mission in Korea (Tilelli, 1999). During an informal meeting with the committee, General Tilelli restated this opinion (Tilelli, 2000). Mines have been used for many years in Korea to defend the Demilitarized Zone and are anticipated to be used extensively if North Korea again attempts to cross the 38th parallel.

Several factors are involved in the decision to use APL in Korea. First, the allies anticipate having to fight with very little warning and being overwhelmingly outnumbered by an enemy seeking to enter Seoul, only 50 miles from the border. Therefore, the faster North Korea can be stopped the better. Second, rugged mountains characterize much of the topography. In this type of terrain, mines are ideal for creating obstacles that can slow a military advance (Troxell, 1999).

General Wesley K. Clark, then CINC of the United States European Command, stated that:

> Self-destructing and self-deactivating APL, and anti-tank (AT)/APL mixed systems constitute a critical force protection and counter-mobility asset. Our field commanders count on these systems to protect the force, influence maneuver, shape the battlespace, and mass combat power for decisive engagement. The requirement for such a capability is increasing in light of evolving and future operational concepts that envision our forces conducting dispersed operations over expanded battlespace. (Clark, 1998)

Concept for Future Joint Operations

The purpose of the *Concept for Future Joint Operations*, issued in May 1997 by the CJCS, was to move the military toward the implementation of *Joint Vision 2010*. The *Concept for Future Joint Operations* is expected to be updated, based on *Joint Vision 2020;* for the present, however, this edition remains authoritative. The *Concept,* which offers a marketplace of ideas and tools for thinking about future operations, identifies the following military-specific trends (CJCS, 1997):

- The proliferation of ballistic and cruise missiles will increase the vulnerability of U.S. and allied forces in theater and jeopardize access to ports and airfields.
- Advanced technology weapons, platforms, and sensors will significantly increase the capabilities of some military forces.

- Microtechnology and biotechnology will create new areas for activity and competition; breakthroughs are likely in the military application of directed energy; and information technology will be vital to military operations.
- Weapons will become more portable and more lethal, and military forces will become more mobile, which will complicate U.S. and allied targeting.
- Some states will rely on asymmetric capabilities (e.g., man-portable air defenses, advanced space capabilities, information operations, landmines, chemical and biological weapons, and terrorism) as substitutes for, or complements to, large conventional forces.

Joint Vision 2010 and the Armed Services

In keeping with their missions and drawing upon their unique capabilities, the Army, Marine Corps, Navy, and Air Force have adapted their force structures, strategies, tactics, people, weapons, and platforms, indeed the way they conduct warfare, to fulfill the broad objectives outlined in *Joint Vision 2010*.[1] The Army and the Marine Corps, the forces that must fight ground wars, rely on landmine capabilities in battlefield environments. The Navy and the Air Force are responsible for the air delivery of certain landmine systems. In addition, each service has a variety of weapon systems capable of destroying enemy tanks and vehicles. If these or future weapons can destroy enemy tanks and vehicles more efficiently than landmines, they might obviate the need for certain landmine systems.

In this section, the strategies developed by each military service for moving toward 2010 are described. The strategies are simple, high-level blueprints, however, and landmines are not specifically mentioned.

U.S. Army

Army Vision 2010 (U.S. Army, 1997a) anticipates the Army's contributions to the operational concepts identified in *Joint Vision 2010*. *Army Vision 2010* is based on the assumption that land forces will exercise direct, continuing, comprehensive control over an area of land, including resources and people, thus solidifying the preliminary advantages achieved by air power. Doctrinally, the Army anticipates being outnumbered and, therefore will depend very heavily on technological superiority.

The geostrategic environment suggests that the Army must be prepared for a range of future missions, which can be categorized into seven general areas: (1) the defense or liberation of territory; (2) a punitive intrusion; (3) containment of a conflict; (4) leverage; (5) reassurance; (6) core security; and (7) humanitarian missions (humanitarian missions that include a range of peace operations are becoming increasingly common) (Rigby, 1999).

The Army is moving toward a combat structure of forces that can be transported and deployed rapidly in times of crisis. The centerpiece of this structure, a system of systems called the Future Combat System, consists of several networked functions, vehicles, and subsystems that could collectively provide overwhelming combat power. Current concepts include an infantry carrier vehicle and robotic vehicles that can provide direct and indirect fire and sensing capabilities. Reconnaissance, surveillance, and target acquisition and command and control will be part of the network (DARPA, 2000a). Scientific and technological research for the Future Combat System may lead to the development of other alternatives to landmines.

U.S. Marine Corps and U.S. Navy

Much of the tactical environment described for the Army will also apply to the Marine Corps. *Operational Maneuver from the Sea* describes an environment in which some operations may require that bases be established ashore. However, most will involve units operating without interruption from ships at sea to their inland objectives. Improvements in the precision of long-range weapons, greater reliance on sea-based fire support, possible reductions in the fuel consumption of military vehicles, and more direct, timely delivery of logistics from sea to users ashore will enhance the Marine Corps' operational maneuvers from the sea. This approach will not be limited to the high end of the spectrum of armed conflict but will be used in a variety of situations, ranging from a struggle against a rising peer superpower to humanitarian relief operations (U.S. Marine Corps, 1997).

With the termination of the Cold War, the Navy abandoned its maritime strategy, based on a war at sea with the Soviet fleet and land-based naval air, and adopted a forward-from-the-sea strategy. The new strategy calls for focusing the power of the fleet and embarked Marines through the littorals and against land targets in regional conflicts (U.S. Navy, 1995).

U.S. Air Force

In response to *Joint Vision 2010,* the Air Force is in the process of changing from the global-reach, global-power strategy of the Cold War to a new strategy called *Global Engagement: A Vision for the 21st Century Air Force* (U.S. Air Force, 1998). The change is based on the conviction that good intelligence, surveillance, and reconnaissance combined with modern aircraft and weaponry will enable the Air Force to find, fix, track, and target anything that moves on the surface of the earth. Information superiority, along with a command and control capability that can coordinate activities and integrate them smoothly with those of the other services, will be an important factor in the Air Force's ability

[1] Changes based on *Joint Vision 2020* are expected but have not been published.

to achieve global-engagement capability. The Air Force's capabilities can be divided into the following core competencies: air and space superiority; global attack; rapid global mobility; precision engagement; information superiority; and agile combat support. As the lead service for space exploration, the Air Force recognizes that space assets will also be vitally important to its own operations, as well as to the other services.

Contrasting Opinions

The contentious nature of APL is reflected in diverse opinions about their use, even within the military. In an open letter to President Clinton published in the *New York Times,* 15 senior, well-respected retired military officers announced their support for a potential ban on APL: "We support such a ban as not only humane, but also militarily responsible" (*New York Times*, 1996).

BENEFITS AND VULNERABILITIES OF NEW TECHNOLOGIES

With no threatening peer competitor to plan for and with the continued rapid emergence of new technologies, particularly information technologies, this would appear to be an opportune time for the United States to make a concerted effort to replace (or at least improve) the systems that currently provide APL functionalities. In addition to retaining the desirable characteristics of APL, new systems should satisfy new requirements, including the capability of distinguishing among friends, foes, and noncombatants rapidly and reliably, easy recovery after hostilities, and environmentally benign effects (see Chapter 3).

Like most innovations, new technologies are bound to have limitations. As weapon systems become more complex, they will also become increasingly vulnerable to breakdowns and to enemy countermeasures. Better C4ISR (command, control, communications, computers, intelligence, surveillance, and reconnaissance) capabilities will entail increased bandwidth, which will also increase the potential for unintended interference among friendly electronic emitters. Improvements in technology should improve U.S. warfighting capabilities; at the same time legacy systems must be able to absorb these advances.

Revolution in Military Affairs

The U.S. military is experiencing what many experts call a "revolution in military affairs" (Krepinevich, 1994). In their classic book *War and Anti-War: Survival at the Dawn of the 21st Century*, futurists Alvin and Heidi Toffler observe that "the way we make wealth is the way we make war—that today's revolutionary changes in business are being mirrored in the world's armies and the future of war itself" (Toffler and Toffler, 1993). Just as the Agrarian Age gave us the hoe and sword, and the Industrial Age gave us mass production and mass destruction, the Information Age will give us the means to fight smarter and more effectively.

The revolution in military affairs is defined as a major change in warfare brought about by the innovative application of technologies that, combined with dramatic changes in military doctrine and operational concepts, will fundamentally alter the character and conduct of operations. Systems, such as the bow and arrow, the rifled gun tube, or aircraft, that use dramatically new technologies can create a major break with the past (Krepinevich, 1994). The new tools invariably affect tactics, operational concepts, and strategies.

Technology, particularly information technology, now defines the possible and is pushing old ideas, values, methods, and organizations into obsolescence (Metz, 2000). Although information technology was used for military purposes prior to 1990 in isolated instances, the beginning of the present revolution in military affairs is usually traced to the Gulf War, when the U.S. public became aware of "smart weapons" and other advances (O'Hanlon, 2000). Since then, the U.S. military has focused more and more on using technology to gain battle space advantages and reduce U.S. casualties.

Many advances in technology will be essential to alternatives to APL, especially in the areas of munitions, information, and communications. The following examples are provided as indicators of where these technologies might take us:

- *Killing Devices.* High-energy explosives that release energy very quickly have improved fragmenting munitions and shaped-charge weapons. Munitions that must penetrate hulls, armor, and other obstructions before they detonate require insensitive, high-energy explosives. The major outstanding issue is achieving a balance between insensitivity and performance (NRC, 1997).
- *Sensors.* Infrared imaging systems have clearly demonstrated their value on the battlefield. Affordable, cooled and uncooled staring focal-plane arrays and associated components that can operate in the mid-wave infrared and long-wave infrared bands will greatly enhance their value. This technology will significantly reduce the costs of operations and provide warfighters with better performing, smaller, lighter infrared imaging systems (Samuels and Supola, 2000).
- *Miniaturization.* Microelectromechanical systems (MEMS) are a revolutionary enabling technology. Embedded into weapon systems, MEMS will provide new levels of situational awareness, information, precision strike capabilities, and new weapons by providing the advantages of small size, low power, low mass, low cost, and high functionality to integrated electromagnetic systems. The primary goal of the DARPA

MEMS program is to develop technology that merges sensing, actuating, and computing into new systems that will increase the perception and performance of weapon systems and the control of battlefield environments (DARPA, 2000b).

- *Platforms.* Advances in the development of unmanned aerial vehicles might enable a platoon pinned down by enemy fire to launch a bird-sized aircraft and use its video camera to look over the horizon, behind buildings, and beyond the range of average eyesight. These micro air vehicles might be able to fly miles from their takeoff point for hours, all the while feeding video images back to ground stations that can use the information to coordinate ground attacks and air strikes (Braham, 1999).
- *Connectivity.* The Army's multifunctional, on-the-move, secure, adaptive, integrated communication project (MOSAIC) will be an energy-efficient, wireless, mobile communications system that provides reach-back and secure networked sensor integration. The open systems architecture will feature increased survivability (Kern, 1999).

Limitations and Vulnerabilities

The seductive promise of advanced technologies should not obscure their vulnerabilities. Future conflicts based on the success of precision engagement, with precision-guided munitions and information dominance, may be advantageous if all combatants are technologically advanced. However, it is somewhat far-fetched if one side relies on less sophisticated small arms and guerrilla tactics. Advanced weapons and smart weapons may not be effective against an adaptive enemy (Scales, 2000).

Even advanced technologies must adhere to the laws of physics. Propulsion systems are generally changing at modest rates. Sensors, however advanced, still have limited abilities to see through many substances. Communications are vulnerable to a variety of attacks, especially to electromagnetic pulse (O'Hanlon, 2000).

The recent air war in the Balkans is an example of mixed success by new technologies. According to press reports, after 78 days and 38,000 combat sorties in Serbia/Kosovo, NATO investigators could only confirm that a dozen Yugoslav tanks had been destroyed, along with 18 other armored vehicles and 20 artillery pieces. Compare this to the original claim that 120 tanks, 220 other armored vehicles, and 450 mortars and artillery pieces were destroyed (*Washington Post*, 2000).

Although we currently have a technological advantage even over our friends and allies, this superiority may not last because the United States is not the only country interested in advanced technologies. For example, a recent book by Chinese military theorists has been described as a Chinese attempt to "explore how technology innovation is setting off a revolution in military tactics, strategy and organization" (U.S. Embassy Beijing, 1999). The widespread availability on the world market of new technologies will certainly neutralize some of the advantages the U.S. military currently enjoys.

Potential Uses of Mine-like Systems

Despite radical changes in advanced weaponry, the fundamentals of land warfare have changed very little since World War II. Therefore, the need for traditional AT mines and APL capabilities is likely to continue. A mine-like system acting as a force multiplier might be very useful to small units facing a numerically superior foe. The sensing and alerting functions currently provided by mines could provide valuable real-time inputs to a tactical information system. In addition to the traditional uses of landmines, the following examples illustrate the potential benefits of mine-like capabilities in future operations.

Seizure of an Airfield

A logistically efficient mine-like system that could be dropped from aircraft and turned on and off using coded transmissions could be deployed in large numbers over and around an airfield during or just prior to an air assault. If the devices could be turned off locally, perhaps within a 50-meter radius of each U.S. soldier, the remainder of the activated munitions would provide a large tactical advantage to the U.S. force. The munitions would warn of approaching enemy troops or tanks, inhibit enemy mobility and reinforcement, and cause some enemy casualties. If the munition sensors were based on radio signals, they could identify friends and foes with passive or semiactive electronic devices.

Building Clearance

Once a building or facility has been cleared of personnel, it would be useful to have a system capable of maintaining the building's cleared status without requiring large numbers of troops. Current mines could be used, but the risk of injury to friendly forces and noncombatants would be high. Future mine-like devices consisting of communicating sensors and nonlethal munitions would be safe for civilians and could automatically warn the tactical information system of intrusions. This mission would require a sensing system capable of detecting and tracking personnel remotely and reporting their movements automatically. Commercial motion sensors could be used for this today, but laser-based or radio-frequency sensors would have longer range and better tracking and discrimination properties. Radio-frequency sensors would have the added advantage of working well in smoke or fog.

Guarding of Stockpiles and Supplies

Systems that could guard stockpiles and supplies, particularly during peace operations, must reliably deter thieves and intruders but must not be lethal to anyone except identified foes. This may require munitions that can give a variety of responses, ranging from warnings (e.g., flashers or sirens) to painful or distasteful events (e.g., sting balls or malodorants). The option for rapid lethal responses when required would also be useful.

Differential Mobility

A remotely deployable, controllable "munition field" equipped with an identification of friend-or-foe capability would allow U.S. soldiers and vehicles to move freely while impeding all others. With this system, a munition field could be placed around and in front of a U.S. position allowing soldiers to withdraw or attack through it, as the situation dictated. The system could even be deactivated long enough to allow an enemy to enter the munition field and then reactivated for an efficient enemy kill. This capability could make the difference between victory and defeat for an initial-entry force because the first units to land on hostile soil are certain to be initially outnumbered by enemy forces.

Border Security

Along borders (e.g., the Demilitarized Zone in Korea and, perhaps, the Kuwaiti-Iraqi border), mines are both deterrents and alerting devices. They can also inflict casualties in the event of a full-scale attack. However, once cleared lanes have been established, current mines become less effective. Future systems could be capable of remaining effective without giving away their positions or being easily deactivated or cleared within a zone. Munitions could be equipped with noncontact sensing and tracking devices and some remote or area response. Remote response could be provided by rapid-response rockets, antipersonnel versions of the Wide Area Munition (WAM), or robotic snipers. Area response could be provided by large fuel and air munitions, either placed in the ground or delivered remotely from a short distance away.

3

Current Uses of Antipersonnel Landmines

Landmines are subtle and much misunderstood weapons. Traditionally they are a means of transforming the terrain to the defender's advantage, rather than providing a definitive barrier. They can inflict casualties but must be covered by fire. They shape the attacker's posture, but do not define the outcome of the battle. They provide economies in defense while imposing attrition on the attacker. They are laid without relish and contemplated with fear. They are simple to lay but remarkably difficult to remove. They are not activated unless an attacker advances, but they do not recognize cease-fires. (Croll, 1998)

This chapter provides a general overview of current U.S. landmine systems, the doctrinal guidance for their use, the capabilities of U.S. APL, and the technologies of current APL. Although this study addresses alternatives to APL, considerable information relating to AT mines is included because APL are often used together with AT mines or in mixed systems that include AT mines. Appendix C provides descriptions of both APL and pertinent AT mines.

DOCTRINAL GUIDANCE FOR USING LANDMINES

The Army, as DOD's executive agent for landmine warfare, is responsible for the doctrine for the use of mines. The Army is also responsible for coordinating these activities with the Marine Corps, Navy, and Air Force. According to doctrine prescribed by the Office of the Joint Chiefs of Staff, landmines are only one component in the overall strategy for constructing barriers and complex obstacles (JCS, 1999). Box 3-1 lists the advantages and disadvantages of using barriers, obstacles, and landmines.

Obstacles can be features of the terrain that impede the mobility of a force. Some obstacles, such as mountains, rivers, railway embankments, and urban areas, exist before the onset of military operations. Military forces create other obstacles to support their operations. Commanders use these obstacles to support their scheme of maneuver. When integrated with maneuver and weapon fire, obstacles can be decisive on the battlefield. Doctrine for some obstacles that rely on a physical object to impede vehicles or dismounted soldiers, such as antitank ditches, wire, road craters, and many types of roadblocks, has not changed since World War II. Because these obstacles do not damage or destroy equipment, or injure or kill soldiers, they are considered to be passive.

Although minefields are also obstacles, they are not passive, and doctrine for mine warfare has changed significantly. Today's mines are different from the mines of the World War II era, which required physical contact and relied on blast effects. Today's mines are triggered not only by pressure, but also by seismic, magnetic, or other advanced fuzes. Mines that self-destruct at preset times give commanders control over how long they remain obstacles. The invention of programmable mines that can recognize and

BOX 3-1
Barriers, Obstacles, and Mines

Advantages

- Provide the capability of inflicting significant damage to equipment and psychological damage, as well as personnel casualties
- Extend, strengthen, and deepen other defensive and offensive measures
- Immobilize the enemy until barriers, obstacles, or minefields can be bypassed, breached, or cleared
- Exploit geographic features
- Free forces for other tasks
- Discern enemy intentions (the commitment of breach assets into a minefield is a detectable indication of intent)
- Create uncertainty for the enemy commander

Disadvantages

- Creation and removal require significant amounts of time, materiel, equipment, and transportation and are manpower intensive and hazardous
- Can be bypassed, breached, or cleared
- Can cause casualties to friendly forces and noncombatants, as well as limit friendly mobility
- Defensive minefields must be rendered safe when they are no longer needed

Source: JCS, 1999.

attack specific types of vehicles will add another dimension to the battlefield (U.S. Army, 1994).

ROLE OF LANDMINES IN WARFARE

From a theoretical point of view, several "principles of war" underpin land combat operations. Landmines could be considered appropriate weapons in the execution of all of these principles but figure most often in the following four: *economy of force* (using all combat power available as effectively as possible; allocating minimum essential combat power to secondary efforts); *security* (never permitting the enemy to acquire an unexpected advantage); *offense* (seizing, retaining and exploiting the initiative); and *maneuver* (placing the enemy in a position of disadvantage through the flexible application of combat power). The use of landmines is scenario dependent and involves trade-offs among the principles of war (U.S. Army, 1993).

On the one hand, mines can contribute to the U.S. military's economy of force and security roles in certain defensive scenarios, especially in sectors that are less important and have been allocated fewer forces and weapons. No doubt, given sufficient forces, it would be more advantageous to have no economy of force sectors in a defensive fight. However, force-multiplier and economy-of-force capabilities are usually necessary. Economy of force does not mean that minefields are unobserved, it just means that the preponderance of combat power is concentrated in another sector of the battlefield, and the economy-of-force sector has to make do with less. APL help defenders do just that.

On the other hand, it is generally accepted that wars are won by attacking the enemy. Mines can play a significant role in causing an enemy to move in directions that are advantageous to friendly forces, allowing them to kill the enemy more efficiently. In some offensive operations, however, U.S. mines previously emplaced during defensive missions that cannot distinguish friend from foe could hinder the execution of rapid offensive maneuvers.

In this report, the committee adopted a long-term view of landmines that took into account the principles of war and a wide range of scenarios, from major theater wars to peace operations. Although this report is focused on the characteristics of APL, APL are only one part of an integrated set of battlefield operating systems designed to give U.S. forces a superior means of fighting and winning an engagement (Bornhoft, 1999). These battlefield operating systems require constant coordination to ensure synergy among intelligence, maneuver, mobility/survivability, fire support, air defense, combat service support, and command and control systems.

The 1998 edition of *Field Manual 20-32* (U.S. Army, 1998b) is the principal U.S. doctrinal basis for the use of minefields; and the 2000 edition of *Field Manual 3-34.2* (U.S. Army, 2000) covers combined arms obstacle integration. Landmines can be used individually to reinforce nonexplosive obstacles or in groups known as minefields (or mined areas in the CCW). A minefield can contain a single type of mine or a combination of types (i.e., APL and AT mines). According to *FM 20-32*, minefields are used for the following purposes:

- Produce a vulnerability on enemy maneuver that can be exploited by friendly forces.
- Cause the enemy to break up his forces.
- Interfere with enemy command and control.
- Inflict damage on enemy personnel and equipment.
- Exploit the capabilities of other weapon systems by delaying enemy forces in an engagement area.
- Protect friendly forces from enemy maneuver and infiltration.

There are four basic types of minefields. *Protective minefields* are used to protect soldiers, equipment, supplies, and facilities from enemy attacks or other threats. *Tactical minefields* directly limit the enemy's movements in a way that gives the defending force a positional advantage. Tactical minefields are designed, sited, and integrated with supporting weapons to produce four tactical effects: disrupting, diverting, fixing, and blocking an enemy. *Nuisance minefields* (another form of tactical minefield) cause enemy forces to move cautiously, thus disrupting, delaying, and sometimes weakening or destroying follow-on forces. *Phony minefields* are used as decoys to deceive the enemy about the exact location of real minefields.

CAPABILITIES OF ANTIPERSONNEL LANDMINES

The committee thought it important to describe the specific capabilities APL bring to a battlefield. A complete treatment of the capabilities, effectiveness, and challenges of APL is beyond the scope of this report, but this synopsis will provide a helpful context for the reader trying to understand this complex issue.

Today's nonself-destructing mines are simple, reliable, low-cost weapons that can operate in all weather conditions, provide deterrence through fear, and can physically delay or kill the enemy. However, they cannot distinguish between friends, foes, and noncombatants, and if they remain active long after the end of hostilities, they can and do kill innocent noncombatants. For these reasons, the U.S. plans to use mines with self-destructing and self-deactivating capabilities that reduce residual explosive hazards on battlefields. The capabilities of APL listed in Box 3-2 were described to the committee during introductory briefings.

Psychological Effects

Among all of the frightening elements on a battlefield, landmines appear to have a unique ability to inspire fear. In a unique study, *The Psychological Effects of Anti-Personnel*

BOX 3-2
Capabilities of Antipersonnel Landmines

- Alert the defender of an enemy approach and effectively delay and/or prevent dismounted attacks.
- Make obstacles and barriers more complex and time consuming to breach.
- Work together with AT mines to create a synergistic effect.
- Protect AT mines from rapid breaching or tampering by the enemy.
- Generate surprise and confusion among enemy forces.
- Cause direct casualties and psychological damage to an attacking enemy.
- Provide rear-area protection for combat support and logistics units.
- Act as a force multiplier for small, light forces and early-entry units.

Source: Biering, 1999.

Landmines: A Standard to Which Alternatives Can Be Compared, Eugenia M. Kolasinski surveyed soldiers who had encountered or used mines in situations other than training. Kolasinski found that APL do not always have significant psychological effects,[1] but when they do, the effect is primarily fear. The fear is most likely based on the types of injuries APL can inflict and the certainty of these injuries if a mine detonates. Three major factors amplify these fears: (1) loss of control, helplessness, and inability to fight back against APL; (2) the perception of risk, which varies by individual and is related to the loss of control; and (3) the high level of uncertainty that continues even after an area appears to be clear of APL (Kolasinski, 1999).

TECHNOLOGIES IN ANTIPERSONNEL LANDMINES

All explosive ordnance contain the same kinds of components: a firing train consisting of a fuze,[2] a detonator, and a main charge. The reaction starts when the fuze activates the detonator (a very small charge of highly sensitive materiel) and progresses to the main charge (larger but less sensitive than the detonator). The type of ordnance is distinguishable by the type of launch or employment that, in turn, defines the shape, body composition, and, to some extent, the configuration of the main charge. Figure 3-1 shows the components of a mine.

Components

The fuze dictates how and when the mine is detonated. The general goal is for the fuze to be safe during handling, storage, transport, and tactical movement; the fuze should be armed only when the munition is used. For instance, a hand-emplaced mine will be fully armed only after being placed in or on the ground, when all safety devices are removed. The complexity of a fuze is as varied as the ordnance to which it is attached. In a mine, the fuze can be mechanical (e.g., a firing pin and a stab detonator), chemical (e.g., acetone that decomposes a plastic disk), electrical (e.g., a battery-operated electrical circuit), or a combination of types. Trip wires or some means of sensing the passage of an individual are common mine fuzes.

The detonator is normally a primary explosive that initiates the booster, which amplifies the detonation and, in turn, initiates the main charge. The body of the mine may be metallic or nonmetallic.

Mine Function

One distinguishing feature of mines is that they lie dormant until a target approaches or contacts them, thus activating the fuze. Another distinguishing characteristic is their instantaneous reaction. Unlike many weapon systems, there is no latency period between when the mine is alerted and when it reacts. In some respects, mines are more precise than other munitions because the action of the target causes them to function. Most mines work in all weather and light conditions. In other weapons, the munition explodes when it arrives at its destination. The main drawback to mines is that they cannot discriminate among friends, foes, and noncombatants.

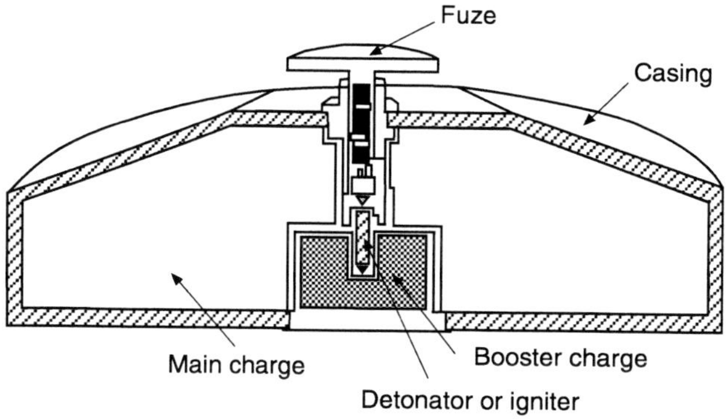

FIGURE 3-1 Mine components.

[1] The effects of APL on soldiers varied significantly. Some soldiers could mitigate their fear by focusing strongly on doing their jobs. Soldiers who had some level of control over where they were going or the types of tasks they were assigned were sometimes less affected.

[2] *Fuse* and *fuze* are not interchangeable terms. A *fuse* is a safety device, that is, a gunpowder-filled cord. A *fuze* is a mechanism that activates a mine or bomb (Croll, 1998).

The necessity of contact or proximity to the target is another distinguishing feature of mines. APL are directly activated by contact, typically the pull of a trip wire. In AT mines, the magnetic field generated by an armored vehicle is sufficient to initiate a fuze designed to fire on magnetic influence. Acoustic signatures can also arm firing circuits. Infrared can be used to detect the presence of a target within the mine's range and fire the main charge. Antihandling devices can be incorporated to preclude easy removal or neutralization.

A mine does not normally explode unless it has been armed. To arm some mines, the user must position the fuze and disengage the safety device (usually by removing a safety pin). The fuze, the initial component in the firing chain, dictates how and when the mine functions. APL are issued with standard fuzes although alternative fuzes may be available. Once armed, mines can be activated in the following ways:

- Pressure can activate the device when a load is placed on it (e.g., contact by walking or driving directly over the mine).
- The release of pressure can set off a mine (e.g., when a pressure-release antihandling firing device attached to the underside is lifted).
- Seismic activity can activate the device when a sensor detects vibrations or movement within the search range.
- Trip wires or break wires can activate a mine if they are disturbed. Trip wires may be either taut (which explode a mine when cut or pulled) or slack (which explode a mine only when pulled).
- Command-detonated mines are activated by a person when he/she detects an enemy in the mine's lethal area. Radio or hard wire can send the signal to fire.

Two types of kill mechanisms are used in APL, blast mechanisms and fragment mechanisms.[3] Blast warheads, which achieve their effects as a result of the detonation of the explosive main charge, can cripple or destroy the foot or leg of a soldier who steps on the mine or burst the tires of a wheeled vehicle that passes over it. Fragmentation warheads disburse metal fragments from the casing or added materials. There are three variations of fragmentation warheads:

- A bounding fragmentation mine projects a canister into the air; the canister explodes and scatters fragments through the lethal area.
- A directional fragmentation mine uses an explosive charge to propel high- velocity fragments over a specific, concentrated area.
- A stake fragmentation mine scatters fragments in a circular pattern.

BOX 3-3
Unexploded Ordnance Hazards

Historically, about 5 percent of explosive ordnance fails to explode on impact. When large quantities of munitions are fired into an area, unexploded ordnance can pose an obstacle to maneuvers, particularly for dismounted troops and unarmored vehicles, as well as for noncombatants after hostilities have ended. At the American Defense Preparedness Association Munitions Executive Summit in Fairfax, Virginia, on September 17, 1996, Paul G. Kaminski, then undersecretary of defense for acquisition and technology, stated the rationale for using self-destructing fuzes for all submunition warheads. "This new fuze will enhance reliability and also provide a self-destruct capability for eliminating the problem of unexploded ordnance on the battlefield" (Kaminski, 1996). This capability will be similar to the self-destructing fuze on U.S. scatterable mines and will provide similar operational and humanitarian benefits.

Self-Destructing and Self-Deactivating Features

Self-destructing and self-deactivating fuzes are essential components of all scatterable mines used by U.S. forces. Self-destructing and self-deactivating APL, used in accordance with international law, do not create a significant humanitarian hazard. The self-destruct feature is designed to explode the mine after a set time has elapsed, thereby eliminating residual effects. Self-destruct times may be as short as four hours or as long as 15 days. The self-deactivating feature uses a battery in the electrical firing system of the mine. When the battery (reserve cell) no longer has enough energy to fire the electrical detonator, the mine becomes inoperable. Self-deactivation normally occurs about 60 days after the mine is armed; at this point the fuze is inoperable. Although the main charge explosive remains, it is relatively insensitive and less hazardous than most other unexploded ordnance. Self-destructing and self-deactivating mines may be the most important contribution to addressing humanitarian concerns about mines.

The self-destructing and self-deactivating capability of today's U.S. scatterable landmines is a desirable operational capability for all nonrecoverable, explosive munitions because they both increase maneuver options and address

[3] Two other types of mines can be used against personnel although they are not normally thought of as APL. Chemical mines dispense chemical agents either by command detonation or vehicle activation (use of these agents is now prohibited by the 1996 Chemical Weapons Convention. Flame mines (flame fougasse) can be fabricated in the field using containers filled with an explosive-fuel mixture activated by an explosive charge. These area weapons are effective against personnel and are normally command detonated.

humanitarian concerns by reducing residual explosive hazards. However, landmines are not the only ordnance that can leave residual unexploded devices on the battlefield. Box 3-3 describes a possible solution to the more widespread problem of unexploded ordnance.

Recommendation. Any nonrecoverable, explosive alternative to antipersonnel landmines should have self-destructing and self-deactivating fuzes to meet operational requirements, address humanitarian concerns, and reduce fratricide among friendly troops. The U.S. government should consider equipping all nonrecoverable explosive munitions with similar technologies.

Nonlethal Weapons

Because the world is so politically unstable, small-scale contingencies and operations other than war may become more common in the future. This type of warfare is likely to occur in areas with significant noncombatant populations where local enemy forces move among noncombatants. Indeed, on occasion, the enemy may use noncombatants as human shields. The United States and its allies will certainly take all prudent steps to avoid causing noncombatant casualties while taking action to eliminate the enemy. Engagements of this nature require a scaled response, which nonlethal weapons can provide.

The United States is experimenting with several types of nonlethal weapons (Fenton, 1999). Nonlethal alternatives to APL would generally replace the fragmentation and blast effects of APL with other capabilities, such as explosively propelled rubber balls, rapidly deployable nets, and electrical charges. Although nonlethal variants by themselves cannot replace APL, they would be useful in certain military operations. With nonlethal variants, U.S. forces could mount a graduated response in situations where the threat were unclear, such as in peace operations, or if large noncombatant populations were in the immediate tactical area. Nonlethal weapons have several advantages: they can be used in a broad variety of circumstances; they can be triggered automatically; and they do not require man-in-the-loop operation to be Ottawa compliant, which can improve the timeliness of a response and lessen the burden on the soldier/operator.

The mission need statements (provided in Appendix G) that describe the military requirements for alternatives to pure APL include nonlethal capabilities as acceptable alternatives. Although the mission need statements for mixed systems did not include nonlethal attributes, the committee decided to include them in the study. Early designs for the Track I NSD-A entailed some nonlethal options, such as nets. Although nonlethal variants by themselves cannot replace APL, for certain military operations they might be useful in support of APL alternatives.

4

Evaluation Methodology

There is no formula or specific set of criteria on which to base a determination of the point at which the putative military utility of AP mines is exceeded by the humanitarian costs of their employment. This is a complex problem involving subjective judgment; for a nation to deny its military the use of a weapon that could assist in prevailing against opposing forces while reducing friendly casualties, there must be a compelling case that the weapon is not indispensable to legitimate military operations and that its utility is clearly outweighed by the resultant killing and wounding of innocent civilians. (Gard, 1999)

The committee considered a large number of potential alternatives to APL. Some are under development as part of the U.S. government's three-track initiative, and some are being developed by other nations. Still others were suggested by nongovernmental organizations and independent scientists committed to the elimination of APL. Members of the committee also suggested several possible alternatives; some are improvements to existing systems and some totally new. This chapter describes the methodology, baseline systems, criteria, and performance factors the committee used to rank potential alternatives.

METHODOLOGY

Information about the alternatives described in this report was gathered over a 10-month period. The sponsor, in meetings with the committee, stressed that all types of alternatives should be considered: materiel and nonmateriel, as well as existing, developmental, conceptual, and notional technologies. Although the committee cast as wide a net as possible, some potential alternatives may have been missed. At the time the data-gathering phase of the study ended, the Track III initiative had reduced the number of concepts under consideration to seven finalists. Three of these were undergoing concept evaluation by industry and were considered competition sensitive; therefore, these were not considered in this study.

Although independent modeling or simulations may have assisted the committee in its evaluations, this was not feasible within the committee's constraints of time and budget. Therefore, the committee developed a score sheet to measure the effectiveness of alternatives. Although the analysis is admittedly subjective, and the results are qualitative rather than quantitative, the scoring was helpful for ranking the alternatives.

The criteria for the score sheet were developed by the entire committee. Committee members with technical expertise provided detailed write-ups of each of the candidates. A subcommittee of technical and military operational experts then met and scored the alternatives against the criteria. The results were presented to the full committee, and adjustments were made, as necessary. The score sheets resulting from these assessments are provided in Chapters 5, 6, and 7.

The committee adopted the terminology used in previous studies by DOD to categorize types of alternatives. *Materiel alternatives* are weapon-like systems; *nonmateriel alternatives* are nonweapon solutions, such as changes in doctrine, organization, concepts, and training or weapons designed for another purpose. Only the materiel alternatives were scored.

BASELINE SYSTEMS

There are many very different types of mines. Mines used solely against a dismounted enemy are so different from the mine systems used against mounted targets that the committee divided its alternatives into two categories: those to be used against a dismounted threat and those to be used against a mounted threat. It then selected two baseline systems against which candidate systems could be measured for all criteria: the M14/16 nonself-destructing mines for use against a dismounted threat and the Volcano system (M87), a mixed system, for use against a mounted threat.

CRITERIA

The committee developed the criteria as a qualitative guide for assessing how well alternatives would fulfill the requirements for APL. These are availability, military effects, humanitarian concerns, overall technical risk, tactics and doctrine, and economic factors. The first three sets of criteria were most important to the committee's overall assessments and recommendations. The committee reiterates that the scores are qualitative and, therefore, somewhat subjective. They were used to help organize the committee's thinking, but not as a basis for pass/fail judgments.

Availability for Implementation

The first and most important criterion was the availability of the system, in keeping with the requirement in the Statement of Task that the committee identify technologies that would be available by 2006. Although 2006 was the critical milestone, many emerging technologies that most likely would not be available until after that date showed considerable promise. Ultimately, the committee divided the potential alternatives into three groups: systems available now (see Appendix C and Chapter 5 for detailed descriptions); systems currently under development that have a reasonable likelihood of being available by 2006 (Chapter 6); and technologies unlikely to be available until after 2006 (Chapter 7).

Military Effects

The second criterion was whether the alternative system could fulfill DOD's requirements (i.e., be militarily effective). The committee's choice of the criterion was neither arbitrary nor capricious. After several attempts to design military criteria for landmines, the committee determined that the capabilities described in the mission need statements[1] were the most complete description available.

A mission need statement is one of the first documents to describe the requirement for a new piece of military equipment. The mission need statements used in this study were validated by the Joint Requirements Oversight Council, a high-level element of the Office of the Joint Chiefs of Staff, and were considered adequate by DOD for industry to use as a basis for designing alternatives. Therefore, the committee considered them to be an accurate reflection of military service requirements. Although the performance standards in these documents are at a higher level than the requirements for technologies described in the study's Statement of Task (i.e., they would perform better than current APL, not similarly), they provide the most complete description of the requirements for mine alternatives known to the committee; an approved operational requirements document was not available. The standards can be read in their entirety in Appendix G.

The numbering of each criterion coincides with the numbers on the score sheets. A score of zero shows that the alternative performed at the same level as the baseline system. A positive number indicates that the alternative performed better than the baseline system. A negative number shows that the alternative did not perform as well as the baseline system.

[1] The two current mission need statements that relate to APL are *Battlefield Shaping and Force Protection against Personnel Threats* (alternatives to APL when used alone) and *Mixed Landmine Systems Alternatives* (alternatives for current mixed systems, combining both APL and AT mines) (see Appendix G for full text).

Against Mounted Threats

The *Mixed Landmine System Alternatives* mission need statement describes requirements for a mixed system for use against a mounted threat (see Appendix G). The committee used the Volcano (M87) as the baseline system against which alternatives were judged.

A1 Enhance effects of close and deep friendly fires
- 0 = same as Volcano
- 1 = greater coverage than Volcano and/or technological improvements
- 2 = man-in-the-loop or improved sensors
- 3 = sensors with battlefield awareness
- –1 = nonlethal component or easily detectable
- –2 = longer sensor-to-shooter cycle
- –3 = longer sensor-to-shooter cycle and a small footprint
- –4 = longer shooter-to-sensor cycle, a small footprint, and may require a cease fire from other weapons to protect the delivery platform

A2 Has multiple methods of delivery
- 0 = at least two methods of delivery (as many or more than Volcano)
- –1 = only one method of delivery

A3 Provides a range of effects that inhibit mounted and dismounted maneuvers
- 0 = similar to Volcano, has both AT and APL components
- 1 = sensor that enables use of something other than an APL mine to counter dismounted targets
- 2 = multiple sensors/highly sophisticated sensors
- –1 = inhibits only personnel or only tanks; nonlethal against personnel
- –2 = nothing in place on ground to stop or slow enemy movement

A4 Resists full spectrum of enemy breach methods, including dismounted methods
- 0 = has AP and AT components
- 1 = has man-in-the-loop to react to visual observation of breach
- 2 = unbreachable without great risk to troops; very difficult to breach
- 3 = multiple sensors, allowing other weapons to be brought to bear; larger area covered; automatically counters breach attempts
- –1 = nonlethal AP components; some AT components have antihandling devices
- –2 = has only AT or only AP
- N/A = no minefield to be breached

A5 Provides early warning of ground attack
- 0 = might provide early warning within normal observation range
- 1 = will provide early warning because of man-in-the-loop (observes ground attack)

2 = will provide reliable early warning via sensors at greater distance to observe attack
–1 = will not provide early warning within normal observation range

A6 Is safe for friendly forces
All alternatives approved in the mission need statements will be as safe as Volcano by doctrine. DOD will not develop a weapon that poses a threat to its own forces.
0 = as safe as Volcano

A7 Is effective in all types of terrain and weather
0 = terrain and weather have no effect
–1 = cannot see target because of weather/terrain; reaction requires sensor input that can be affected by weather/terrain

A8 Poses minimal residual hazard to own forces and noncombatants after military conflicts
0 = can self-destruct/self-deactivate; no ground munitions to explode (munitions have exploded in the air or on contact with the ground)
1 = can be command detonated; is strictly nonlethal

A9 Is difficult to detect by enemy forces
0 = same as Volcano; may be visible when shot off but are not easily visible on ground
1 = smaller and less easily seen than Volcano
2 = nothing to be detected on the ground
–1 = bigger, easier to see; sends a signal that can be detected
–2 = visible and audible by the enemy before breaching

A10 Poses minimal risk of fratricide
0 = has an APL component
1 = has no APL component; has nonlethal AP component
2 = man-in-the-loop allows identification of friend or foe before detonation; more accurate minefield location
3 = sensors tell own troops where munitions are; munitions can be turned off when own troops approaching; is strictly nonlethal

A11 Effects modifiable for mounted and/or dismounted threat
0 = once in place munitions are either AT or AP
1 = has man-in-the-loop, therefore other systems can be used
2 = more sensors allow several options for bringing other weapons to bear

A12 Has controllable activation/deactivation and duration before and after installation.
0 = set before emplacement, cannot be changed after
1 = set at time of use; man-in-the-loop
2 = can do many things with munitions (turn on/off; command destruct)

A13 Is effective in nuclear, chemical, and biological environments
0 = not affected (same as Volcano)
–1 = sensors or communications may be affected by electromagnetic pulse

A14 Is easy and efficient to distribute
0 = remotely delivered; AT and AP components delivered together
–1 = emplaced by hand; components delivered separately (i.e., requires two rounds)
–2 = more components require hand emplacement than Volcano

Against Dismounted Threats

The *Battlefield Shaping and Force Protection Against Personnel Threats* mission need statement describes the capabilities required for protection against a dismounted threat (see Appendix G). The committee used the M14/M16 as the baseline system for evaluating these alternatives.

A15 Can delay, disrupt, and/or canalize enemy movement/maneuvers.
0 = same as M14/M16
1 = greater area covered; greater range; selection of when to detonate; line of sight
2 = detection of enemy forces further out via sensors allow time for use of other weapons
–1 = nonlethal with strong effects that might affect enemy movement
–2 = nonlethal with weaker effects not likely to affect enemy movement

A16 Denies enemy access to terrain/facilities (including short-term/long-term deterrent for boundaries and DMZ areas)
0 = same as M14/M16
1 = greater area covered; greater range; selection of when to detonate; line of sight required
2 = detection of enemy forces further out via sensors, allowing time for use of other weapons
–1 = nonlethal, with strong effects that might affect enemy movement
–2 = nonlethal, with weaker effects that are not likely to affect enemy movement

A17 Enhances effects of friendly force weapons, obstacles, and munitions (including AT mines)
0 = same as M14/M16
1 = greater coverage; man-in-the-loop
2 = better sensors
–1 = nonlethal

A18 Generates exploitable delays and opportunities (fixes or contains enemy)

- 0 = same as M14/M16
- 1 = greater area covered; greater range; selection of when to detonate; line of sight required
- 2 = detection of enemy forces further out via sensors allowing time for use of other weapons
- −1 = nonlethal, with strong effects that might affect enemy movement
- −2 = nonlethal, with weaker effects that are not likely to affect enemy movement

A19 Generates detection, alert, and/or early warning
- 0 = makes a noise, alerting friendly troops to enemy intrusion
- 1 = sensors alert to enemy intrusion
- 2 = more sophisticated sensors alerting of intrusion at greater distance (sooner)
- −1 = no noise is made autonomously; line of sight required; difficult to cover blind spots

A20 Facilitates classification of target
- 0 = no classification
- 1 = man-in-the-loop
- 2 = man-in-the-loop with sophisticated sensor

A21 Produces desired effects on enemy forces (nonlethal to lethal)
- 0 = solely lethal or nonlethal; no munition
- 1 = permits choice of munition (lethal or nonlethal)
- 2 = has inherent selection of effects, from lethal to nonlethal

A22 Reduces casualties/risk for U.S. and/or allied forces
- 0 = same as M14/M16
- 1 = self-destruct/self-deactivating or man-in-the-loop
- 2 = nonlethal; more positive identification at greater ranges

A23 Deters pursuit to facilitate breaking of contact under pressure
- 0 = no man-in-the-loop, weapon does not require operator
- −1 = man-in-the-loop; nonlethal

Humanitarian Concerns

The third criterion was fulfillment of the humanitarian intent of international agreements, such as the CCW and the Ottawa Convention. Although compliance with these agreements was not specifically included in the Statement of Task, the committee considered it implicit in the rationale for this study. Senator Leahy, a strong supporter of demining, providing aid for victims of mines, and the Ottawa Convention, introduced legislation requesting this study (Rieser, 1999). Because humanitarian concerns were clearly one impetus for this study, the committee used those elements from the CCW and the Ottawa Convention that addressed the post conflict hazard of mines and possible injuries to noncombatants as humanitarian criteria. The five characteristics listed below, derived from the CCW and the Ottawa Convention, were applied to alternatives used to counter mounted and dismounted targets.

Against Mounted Threats

B1 Presents postconflict residual hazard (CCW)
- 0 = no self-destruct/self-deactivation mechanism; munitions persist on ground
- 1 = has self-destruct/self-deactivation; easily removable; purely nonlethal; no munitions persisting on ground

B2 Location can be recorded (CCW)
- 0 = approximate locations are known
- 1 = more precise location known
- 2 = will self-record where they are
- 3 = no munition

B3 Detectable to facilitate removal (CCW)
- 0 = sufficient metal content for detection
- 1 = no munitions

B4 Discriminates between combatants and civilians (Ottawa)
- 0 = has no means of discrimination
- 1 = less than positive target identification (i.e., with a nonvisual sensor); observed fire
- 2 = discriminates by man-in-the-loop, visual identification; has nonlethal AP component

B5 Does not explode on presence, proximity, or contact of a person (Ottawa)
- 0 = explodes on presence, proximity, or contact
- 1 = does not explode on presence, proximity, or contact; is nonlethal

Against Dismounted Threats

B1 Presents residual hazard (CCW)
- 0 = no self-destruct/self-deactivation mechanism
- 1 = is self-removing (via self-destruction/self-deactivation) or nonhazardous after war; is very easily removable; is strictly nonlethal

B2 Location can be recorded (CCW)
- 0 = is hand emplaced
- 1 = has man-in-the-loop or sensor

B3 Detectable to facilitate removal (CCW)
- 0 = sufficient metal content for detection
- 1 = no munitions

B4 Discriminates between combatants and civilians (Ottawa)
- 0 = has no means of discrimination
- 1 = less than positive target identification (i.e., with a nonvisual sensor)
- 2 = discriminates by man-in-the-loop, visual identification

B5 Does not explode on presence, proximity, or contact of a person (Ottawa)
- 0 = explodes on presence, proximity, or contact
- 1 = does not explode on presence, proximity, or contact; is strictly nonlethal

Overall Technical Risk

In assessing overall technical risk, the committee assigned the highest score to systems already in production and progressively lower scores to systems with technologies beyond the state of the art. This very simple scoring was helpful for assessing the more "futuristic" alternatives.

C0
- 0 = in production
- −1 = capability has been demonstrated
- −2 = capability not prototyped, but uses technology within the state of the art
- −3 = technology is beyond the state of the art

Changes in Tactics and Doctrine

The consideration of tactics and operational concepts was specified in the Statement of Task. This criterion was given a simple yes or no on the score sheet to indicate whether changes in tactics or doctrine would be necessary to implement the alternative (see **D0** on score sheet).

Cost

Because cost[2] was not included in the Statement of Task and because the determinations by the committee were only estimates, scores for this criterion only indicate the committee's evaluation process. The following costs were considered for each alternative:

E1 Research and development
- 0 = weapon already exists
- −1 = known, funded cost; relatively low cost; prototype exists or is technologically easy to achieve
- −2 = no prototype, but elements of the system exist; technology appears to be straightforward
- −3 = conceptual stage; requires a technology breakthrough

E2 Procurement cost
- 0 = production base exists/has been produced
- −1 = production capability exists, anticipated product costs are low
- −2 = requires complex manufacturing process and/or high product cost
- −3 = manufacturing processes are conceptual, costs unknown

[2] Despite the sponsor's admonition that cost not be considered a factor, it was included because the committee believed it might be a consideration for the future. However, cost estimates were rudimentary.

5

Alternatives Available Today

> To the military, mines are an extremely cost-effective method of denying ground; so much so that extensive studies undertaken in the U.S. have so far failed to find any practical substitute that could be used if a ban were implemented. (Jane's, 1997)

OVERVIEW

The committee first considered whether nonmateriel alternatives, such as changes in tactics and/or operational concepts, could fully compensate for the elimination of APL. Because these alternatives did not lend themselves to the type of scoring used for materiel solutions, the committee simply described possible changes. The committee then focused on materiel solutions (i.e., weapons and weapon systems), including the four existing mine systems that could be considered Ottawa compliant and five precision-guided weapons. In general, the most critical limitation of non-mine alternatives that depend on direct or indirect support weapon systems is the lack of certainty of their availability at precisely the time that they would be required. None of the alternatives can replicate all of the dimensions of the instantaneous response of APL.

NONMATERIEL ALTERNATIVES

The committee believes that DOD has looked to the National Academy of Sciences for advice on *technical solutions* to the landmine issue instead of the many military institutions fully capable of advising DOD on operational concepts and tactics. Therefore, the committee provides only a cursory evaluation of operational concepts and tactics.

The nonmateriel alternatives to landmines have been suggested by various sources, including committee members. First, the committee rejected the obvious option of simply eliminating APL and accepting additional casualties. A variety of operational concepts and tactics that might be effective were considered. Because of the many variables, such as mission, threat, weather, terrain, and other conditions in which APL could be employed, these nonmateriel alternatives were only addressed in a general way.

Operational Concepts

In its simplest form, an APL consists of a sensor (a target-activated fuze), rudimentary communications (e.g., noise), and an instantaneous kill mechanism. Other sensors and kill mechanisms in the U.S. inventory might be used to achieve similar results. The committee considered the following operational concepts as possible alternatives:

- Use mechanical ground systems, such as trip flares and improvised noisemakers. (Disadvantage: not linked to instantaneous lethal mechanisms)
- Use electronic ground systems, such as the remotely monitored battlefield sensor system (REMBASS), ground-based portable radars. (Disadvantages: not generally available at the small combat unit level; heavier than current APL; no delay; no enemy casualties without linked kill mechanisms)
- Provide additional human systems, such as armed soldiers and more effective equipment, including binoculars, night-vision devices, or other capabilities; more deep-reconnaissance/surveillance units in conventional or special operations force organizations. (Disadvantage: increased soldier vulnerability because reaction time can be dramatically slowed by incoming enemy fire, fatigue, weather, darkness, and other conditions)
- Employ animal systems, such as dogs or geese. (Disadvantages: noisy; high logistical cost for animal support)
- Call upon airborne systems, such as unmanned aerial vehicles, helicopters, fixed-wing aircraft, joint surveillance target attack radar system (JSTARS), and satellites. (Disadvantages: increased time between detection of an enemy and receipt of information by the unit

on the ground; are frequently unavailable to small units).

Tactics

The commander on the ground is responsible for accomplishing the unit mission by ensuring that subordinate units or troops use all weapons in a way that exploits the unique conditions of enemy, terrain, weather, and light. A commander might tactically employ soldiers, sensors, weapons, and units in the following ways to provide similar advantages as APL:

- Use more forward reconnaissance (e.g., additional soldiers, ground sensors, and aerial sensors) to determine the enemy's location earlier in the decision cycle so continuous and sequential destruction can be inflicted as the enemy comes into the range of each available indirect-fire system, direct-fire system, and mine-like device. (Disadvantages: requires additional military manpower; greater likelihood of high U.S. casualties)
- Use more soldiers, weapons, or units in a given battlefield area to increase firepower advantage on a given piece of terrain and increase the likelihood of slowing or defeating the enemy. (Disadvantages: requires additional military manpower; greater likelihood of high U.S. casualties)
- Provide small, lightweight containers of contingency sensors and / or weapons that can be moved quickly by ground or air to the position of a small unit. The items in the containers would be tailored for local conditions and could include any combination of night-vision devices, ground sensors, Claymores, grenade launchers, machine guns, hand-held mortars, ammunition, and nonlethal munitions. (Disadvantages: uncertainty that container would be available when needed; additional training required to teach soldiers to use a range of sensors and weapons not normally available to them)
- Employ AT mines "just-in-time" to support maneuver. Conduct a thorough terrain/enemy analysis, make detailed fire plans, and establish priority of fires so that AT mines could be delivered just in time (within minutes rather than hours) to support maneuver. For example, a dangerous enemy avenue into an advancing friendly force's flank could be closed with AT mines delivered remotely just prior to friendly force arrival. This would minimize the enemy's ability to find the scattered minefield, and the passing friendly forces would be able to cover the minefield with real-time observation, direct fire, and indirect fire. (Disadvantage: uncertainty that a dedicated delivery means, such as artillery, would be immediately available)
- Employ remotely delivered AT mines in greater numbers, over greater areas, with more rapid reseeding rates. The larger the minefield the more difficult it may be for the enemy to bypass it (going around a minefield is usually the simplest countermeasure but often leads to a kill zone); a larger minefield is likely to require more time to breach with mine plows or more specialized armored breaching vehicles. As noted in Appendix D, APL are used to slow dismounted breaches of AT minefields. By reseeding existing AT minefields with additional remotely delivered AT mines, both mounted and dismounted breach attempts could be slowed as lanes thought to be passable would have to be recleared. Ideally, reseeding would be accomplished under real-time direction from a ground observer, a manned aircraft, or an unmanned aerial vehicle sensor. Otherwise, high-priority, remotely delivered minefields could be periodically reseeded as the tactical situation required. (Disadvantages: requires additional delivery means, mines, and military personnel)

The effectiveness of any of the tactical approaches listed above would greatly depend on the mission, the situation, and the force structure. Furthermore, history has shown that when one side changes tactics, the other side makes counter-changes. On the battlefield, tactics evolve, sometimes radically. Even though APL are rarely decisive on the battlefield, they do provide a commander with one more capability to shape the battle space, tailor his tactics, and enhance the effects of other more decisive systems. Therefore, the tactical approaches listed above might have a delaying effect but, either singly or in combination, they could not replicate the instantaneous lethality of APL on a precise point on the battlefield.

MATERIEL ALTERNATIVES

Materiel alternatives to APL are likely to consist of a combination of sensor, communication links, and lethal or nonlethal munitions. The committee carefully evaluated technologies in each of these categories in terms of the fundamental problem of current mines—they cannot distinguish between friend and foe. Although significant efforts have been devoted to existing and future communications and munition technologies, ground-based sensors cannot discriminate rapidly and accurately between types of soldiers and /or noncombatants. In the committee's opinion, the development of long-lasting, accurate, all-weather capable, low-power ground sensors may be key to the creation of the most flexible and militarily effective alternatives to APL.

APL have two missions—to kill dismounted targets and to protect AT mines from being breached; the latter is typically accomplished by mixed systems. An alternative to APL in these mixed systems could either (1) remove the APL and use only AT mines equipped with antihandling devices or (2) use other weapons designed for non-mine missions that can destroy tanks and other heavy vehicles without the risk of breaching. For this reason, some of the alternatives considered are AT mines or weapons.

Table 5-1 shows existing systems that would be compliant with the Ottawa Convention. The table also describes their principal characteristics. Following the table are full descriptions of each alternative. The descriptions are followed by brief assessments and tables measuring alternatives are measured against the criteria described in Chapter 4.

For Use Against Dismounted Threats

Claymore (M18)

Source: Mine/Countermine Operations, FM 20-32 (U.S. Army, 1998b)

The M18 series, or Claymore mine, is a nonself-destructing directional fragmentation mine detonated by 682 grams of composition C4. The Claymore projects 700 steel balls in a fan-shaped pattern in a 60 degree arc to a maximum height of 2 meters. The M18 can be activated in the command-detonation mode by an electric blasting cap inserted into the detonator well. The mine body is nonmetallic, and the steel balls are cast in the same composition as the front part of the mine. The lethal radius extends to 100 meters forward and 16 meters to the rear. The Claymore mine is issued with an electric initiation system (hand generator, wire, and electric blasting cap) and is now doctrinally detonated by command, although so-called "mechanical ambushes," using trip-wires, were common during the Vietnam War. Amended Protocol II to the CCW imposes some restrictions on the use of trip wires to detonate the Claymore; the Ottawa Convention prohibits the use of trip wires with all mines. The command-detonated Claymore does not use a trip wire and is, therefore, permitted by both the CCW and the Ottawa Convention.

For Use Against Mounted Threats

Volcano (M87A1)

Source: Mine/Countermine Operations, FM 20-32 (U.S. Army, 1998b)

The Volcano system (M87) replaced the older M56 helicopter-delivered AT mine system and the GEMSS/Flipper system. The M87 is a mixed system, and the M87A1 is a pure AT system. Volcano dispensers can be mounted on several tracked or wheeled vehicles or on the UH-60 Blackhawk helicopter. The system is made up of the launcher rack and dispenser-control unit, vehicle-specific mounting hardware, and mine canisters, each of which holds six mines. A completely loaded dispenser holds 160 canisters, or 960 mines. The mines are placed in a uniform density of approximately one mine per linear meter over a linear distance of one kilometer. Self-destruct times of 4 hours, 48 hours, or 15 days are set at launch time. The Volcano system can be used to emplace protective and tactical minefields anywhere on the battlefield reachable by the dispensing vehicles.

Remote Antiarmor Mine System (RAAMS)

Source: Mine/Countermine Operations, FM 20-32 (U.S. Army, 1998b)

The Remote Antiarmor Mine System (RAAMS) consists of a 155-millimeter (mm) howitzer projectile containing nine AT mines. RAAMS is usually used with the ADAM

TABLE 5-1 Alternatives Available Today

System Name	APL/ AT/ Mixed Non-Mine	Self-destructing/ Self-deactivating	Lethal/ Nonlethal	Ottawa Compliant[a]	Dismounted Enemy		Mounted Enemy	
					Remotely Delivered	Hand Emplaced	Remotely Delivered	Hand Emplaced
Claymore (M18)	APL	N	L	Y		X		
Volcano (M87A1)	AT	Y	L	Y			X	
Remote Antiarmor Mine System (RAAMS)	AT	Y	L	Y			X	
Hornet/Wide Area Munition (WAM)	AT	Y	L	Y				X
Maverick (AGM-65)	n/m	N	L	Y			X	
Longbow Hellfire (AGM-114)	n/m	N	L	Y			X	
Sensor Fuzed Weapon (SFW)	n/m	N	L	Y			X	
Sense and Destroy Armor Munition (SADARM)	n/m	N	L	Y			X	
Brilliant Antiarmor (BAT) Submunition	n/m	N	L	Y			X	

[a]The committee used the definition found in the Ottawa Convention to determine whether a system would be Ottawa compliant.

(Area-Denial Artillery Munition), but can also be used alone. Each mine in the projectile is a right-circular cylinder, 11.25 centimeters (cm) in diameter and 7.5-cm high, weighing 1.8 kilograms and containing about 0.78 kilograms of pressed explosive in the main charge. RAAMS have factory-set self-destruct times of either 4 or 48 hours. About 20 percent have an antihandling feature that causes them to explode when tilted at an angle of 17 degrees or more. They are designed to destroy mounted targets by perforating the underside of the vehicle. The mines have magnetic-influence fuzes designed to stop mounted vehicles when they are detonated, as intended, between the tracks; if they are detonated directly under a track, they may not stop the vehicle.

Hornet/Wide Area Munition (WAM)

Source: Mine/Countermine Operations, FM 20-32 (U.S. Army, 1998b)

The Hornet/WAM is an autonomous hand-emplaced AT mine. It weighs about 16 kilograms, is about 36 centimeters high and about 23 centimeters in diameter. After it is emplaced and armed, seismic sensors can detect movement and alert the mine to turn on its acoustic sensors that detect and classify a target. If an armored target approaches within 100 meters, a small submunition with an infrared sensor is launched over the target and fires an explosively formed projectile down into the engine compartment. The Hornet/WAM is designed to operate for 30 days after arming and then to self-destruct. It has an antihandling feature that causes the mine to detonate when disturbed. The submunition is similar to those in the Air Force Sensor Fused Weapon. Originally, the Hornet/WAM was to be dispensed from a Volcano launcher and/or from a deep-attack asset, such as a multiple launch rocket system or a tactical missile system. Subsequent evolutions of requirements and the exigencies of the development program have precluded these options. For the foreseeable future, the Hornet/WAM will be hand emplaced, which sharply limits its utility.

Maverick Air-to-Ground Missile (AGM-65)

Source: Vietnam Veterans of America Foundation

Called to the attention of the committee by the Vietnam Veterans of America Foundation, the Maverick (AGM-65) is a tactical guided missile designed for close air support, interdiction, and defense suppression (Deagle, 2000). Maverick provides stand-off capability and has a high probability of striking a wide range of tactical targets, including armor, air defenses, ships, transportation equipment, and fuel storage facilities. Because Maverick has a modular design, different combinations of guidance packages and warheads can be attached to the rocket motor section to produce different weapons.

Maverick has three different seekers and two different warheads. The solid-rocket motor propulsion section is common to all variants. The seeker options are electro-optical imaging, infrared imaging, or a laser guidance package. The warhead, either a 56.7 kilogram shaped charge or a 136.08-kilogram penetrator, is in the missile's center section. A contact fuze in the nose fires the shaped-charge warhead. The penetrator uses a delayed fuze, allowing the warhead to penetrate the target by kinetic energy before firing, which is very effective against large, hard targets.

Maverick has a cylindrical body with long-chord delta wings and tail control surfaces mounted close to the trailing edge of the wing of the aircraft using it. Mavericks can be launched from high altitudes to tree top level. The A-10, F-15E, and F-16 aircraft can carry as many as six Mavericks allowing the pilot to engage several targets on one mission. The missile also has "launch-and-leave" capability that enables a pilot to fire it and immediately take evasive action or attack another target as the missile guides itself to the target. Tactical employment of Maverick is fully covered in doctrinal manuals. This weapon is in production and requires no further research and development.

Advantages

- Complies with CCW Amended Protocol II and the Ottawa Convention.
- Can destroy targets from a stand-off platform, which has a psychological impact on an enemy force.
- Is effective against all land-combat vehicles, including heavy armor.

Disadvantages

- Requires a target acquisition capability to locate engagement areas.
- Requires an expensive, complicated launch platform (fixed-wing tactical aircraft).
- Resupply requires extensive lift and transportation capability.
- Uncertainty about the launch platform's ability to support additional tactical missions.
- Because of the delivery system, it is difficult to use close to friendly troops.
- Provides no protection from dismounted troops.
- Although compliant with Ottawa, Maverick may cause serious collateral damage.

Longbow Hellfire Air-to-Ground Missile (AGM-114)

Source: Vietnam Veterans of America Foundation

Called to the attention of the committee by the Vietnam Veterans of America Foundation, the Longbow Hellfire air-to-ground missile (AGM-114) can engage both moving and stationary vehicles and provide an adverse-weather, fire-and-forget, heavy-antiarmor capability for attack helicopters (Deagle, 2000). The Longbow program also includes a fire-control radar system and numerous modifications to the helicopter. The fire-control radar system can locate, classify, and prioritize targets for the Longbow Hellfire missile.

The first three generations of Hellfire missiles use a laser seeker. The fourth generation, Longbow Hellfire, is a fire-and-forget version that uses an active radio-frequency seeker operating in the millimeter-wave frequency band and has a dual tandem warhead designed to defeat reactive armor. Either the Apache Attack Helicopter's (AH-64D) fire-control radar or a laser designator may identify targets. Integration of the Longbow into the entire fleet of Apache attack helicopters and into one-third of the Comanche reconnaissance helicopter fleet is planned. Tactical employment is fully covered by doctrinal manuals. Research and development of this program is supported and funded as an Army missile program.

Advantages

- Complies with CCW Amended Protocol II and the Ottawa Convention.
- Can lock on and destroy targets on the battlefield, which has a psychological impact on an enemy force.
- Can operate in adverse weather (rain, snow, fog, smoke, and battlefield obscurants).
- Has good survivability.
- Has fire-and-forget guidance.
- Can be reprogrammed and adapted to changing threats and mission requirements.
- Has millimeter-wave countermeasures, which allows the munition to be launched and then remasked, minimizing its exposure to enemy fire.
- Is capable of defeating reactive armor configurations.

Disadvantages

- Requires a target acquisition capability to locate engagement areas.
- Requires an expensive, complicated launch platform (AH-64).
- Resupply requires extensive lift and transportation capability.
- Additional tactical missions for the system may require procurement of additional missiles.
- Uncertainty about the launch platform's ability to support additional tactical missions.
- Provides no protection from dismounted attack.

Sensor Fuzed Weapon (SFW)

Source: Vietnam Veterans of America Foundation and Textron Systems

Called to the attention of the committee by the Vietnam Veterans of America Foundation, with system information from Textron Systems, the Sensor Fuzed Weapon (SFW) is an unpowered, top-attack, wide-area cluster munition designed to achieve multiple kills per aircraft pass against enemy moving and stationary land-combat vehicles (Deagle, 2000; Gard, 1999; telephone conversation between J. Johnson, program manager, Textron Defense Systems and R. Johnson, committee member, April 3, 2000). SFW was specifically designed to defeat multiple targets with a single weapon. The SFW's tactical munition dispenser weighs 454 kilograms and houses 10 submunitions, each of which has four target-sensing, armor-penetrating warheads (a total of 40 lethal warheads per weapon).

After deployment, the tactical munition dispenser opens and dispenses the 10 submunitions that are stabilized with parachutes. At a preset altitude sensed by a radar altimeter, a rocket motor fires to spin the submunition and initiate an ascent. It then releases its four warheads, which are lofted over the target area. The warhead's sensor detects a vehicle's infrared signature and an explosively formed penetrator fires at the heat source. If no target is detected after a period of time, the warheads automatically detonate to limit hazardous residue. The baseline weapon has a time-out, self-destruct feature that limits the dud rate to one in 40. The planned improvement of the SFW provides a redundant self-destruct feature that will essentially eliminate battlefield duds. One weapon can effectively neutralize moving and stationary land-combat vehicles within a 15-acre (60,000 square meters or 6 hectares) area.

The SFW can be deployed from operational U.S. or NATO tactical aircraft, such as the F-16, which can carry four munitions. The weapon can be delivered in all weather conditions, day or night. With the 2001 addition of the wind-corrected munition dispenser tail kit, the SFW will be much more accurate.

Tactical use of the SFW is fully covered in doctrinal manuals. This program is in full rate production and will not require additional research or development.

Advantages

- Complies with CCW Amended Protocol II and the Ottawa Convention.
- Autonomously finds and destroys targets on the battlefield, which has a psychological impact on an enemy force.
- Certified on all U.S. bombers and fighter aircraft.
- Provides accurate delivery at all altitudes.
- Is effective against all land-combat vehicles, including heavy armor.
- Has multiple kills per weapon and wide area coverage.

Disadvantages

- Requires a target acquisition capability to locate engagement areas.
- Once launched, the employment time window is limited to seconds (the flight time of the submunition).

- Requires an expensive, complicated launch platform (fixed-wing tactical aircraft).
- Resupply requires extensive lift and transportation capability.
- Because of the delivery system, it is difficult to use close to friendly troops.
- May not be the weapon of choice against targets more than 200 meters apart.
- Not used in conflicts where the rules of engagement require delivery above 4500 meters, when wind-corrected munition dispenser is not available and collateral damage is a concern.
- Although compliant with Ottawa, the SFW could result in increased collateral damage because of the large kill zone.
- Not considered a stand-off weapon (although use of wind-corrected munition dispenser from high altitude can provide horizontal stand-off equal to about 1.5 times the altitude).
- Provides no protection from dismounted attack.

Sense and Destroy Armor Munition (SADARM)

Source: Vietnam Veterans of America Foundation and the Office of the Project Manager-SADARM

The Sense and Destroy Armor Munition (SADARM) was brought to the attention of the committee by the Vietnam Veterans of America Foundation, the system information was validated by the Office of the Project Manager, SADARM (Deagle, 2000; Gard, 1999; telephone conversation between B. Demassi, deputy project director, SADARM, and R. Johnson, committee member, April 3, 2000). The U.S. Army's SADARM was developed to provide an autonomous, counterbattery capability to indirect-fire artillery units. SADARM is a fire-and-forget, multisensor, smart munition designed to detect and destroy armored vehicles. It is fired principally from self-propelled artillery. SADARM uses three sensors (active, passive millimeter-wave, and infrared) to detect armored vehicles in any weather and destroy them from above. The SADARM submunitions are first delivered to the target array by artillery projectiles; each projectile contains two submunitions. Following expulsion from the projectile, the submunitions are dispensed and decelerate to orient themselves. When the first parachute is deployed, the submunitions scan and process the data. Concurrently, the millimeter-wave sensor detects altitude. When the second parachute is deployed, the warhead is armed. When the target is detected, the warhead fires and the target is defeated by an explosively formed penetrator fired into the top of the target. SADARM's range is approximately 20 kilometers and it can be fired from standard 155-mm tube artillery. Tactical employment is fully covered in doctrinal manuals. Research and development is supported and funded as an Army artillery program.

Advantages

- Complies with CCW Amended Protocol II and the Ottawa Convention.
- Autonomously finds and destroys targets on the battlefield, which has a psychological impact on an enemy force.
- Can penetrate all known top armor.

Disadvantages

- Requires a target acquisition capability to locate engagement areas.
- Once launched, the employment time window is limited to seconds (the flight time of the submunition).
- Resupply requires extensive lift and transportation capability.
- Additional tactical missions for the system may require procurement of additional projectiles (and submunitions).
- Because of the delivery system, it is difficult to use close to friendly troops.
- Although compliant with Ottawa, SADARM may cause serious collateral damage.

Brilliant Antiarmor (BAT) Submunition

Source: Vietnam Veterans of America Foundation and Office of the Project Manager, Brilliant Antiarmor Submunition

The Brilliant Antiarmor (BAT) Submunition was brought to the attention of the committee by the Vietnam Veterans of America Foundation (Deagle, 2000); system information was provided by the Office of the Project Manager, Brilliant Antiarmor Submunition (telephone conversation between D. Pinkston, Office of the Project Manager, BAT, and R. Johnson, committee member, April 3, 2000). The BAT submunition uses passive acoustic and infrared sensors to find, attack, and destroy moving tanks and other armored vehicles deep in enemy territory. It is called "brilliant" because it can seek, identify, and destroy armored targets autonomously. An unpowered, aerodynamically stable glider, BAT can operate day or night and in all weather conditions. After separation from the launch missile, BAT's acoustic sensors, working in combination with a high-speed onboard computer, steer it toward the sound of the target vehicles. The infrared sensor then guides the weapon in for the terminal attack phase. BAT destroys enemy targets with a hit-to-kill, conventional, shaped-charge warhead.

BAT is 91.44 centimeters long, 13.97 centimeters in diameter with wings folded, and weighs 19.96 kilograms. The preplanned product improvement BAT retains the same physical characteristics of BAT and has a new dual-mode imaging infrared / millimeter wave seeker. This upgraded terminal seeker adds capability against stationary targets and is more robust in adverse weather and against countermeasures. The target set of the improved BAT includes missiles and rocket launchers. Because the operational software is downloaded rather than resident inside BAT, additional targets can be programmed without hardware modifications.

BAT is designed to be carried by the tactical missile system for "many-on-many" attacks against massed moving armor. It is also designed to be compatible with other delivery vehicles, such as the tactical munition dispenser, which allows it to be dispensed from a jet aircraft and cruise and ballistic missiles, such as the Tomahawk and the Multiple Launch Rocket System (MLRS).

Tactical employment of BAT will be fully covered in doctrinal manuals. Research and development is supported and funded as an Army artillery program.

Advantages

- Complies with CCW Amended Protocol II and the Ottawa Convention.
- Low collateral damage and stand-off attack capability.
- Future costs are only for maintenance and storage.
- Autonomously finds and destroys targets on the battlefield, which will have a psychological impact on an enemy force.

Disadvantages

- Requires a target acquisition capability to locate engagement areas.
- Once launched, the time window is limited to seconds (BAT's flight time).
- Requires an expensive, complicated, scarce launch vehicle, such as the MLRS combined with the Tactical Missile System (TACMS).
- Resupply requires extensive lift and transportation capability.
- Additional tactical missions may require procurement of additional missiles (and submunitions).
- Launch platform may not be able to support additional tactical missions.
- Because of the delivery system, it is difficult to use close to friendly troops.
- Provides no protection from dismounted attack.

COMMITTEE ASSESSMENTS

Nonmateriel Solutions

The committee could not identify alternative tactics or operational concepts that could, on their own, provide tactical advantages to U.S. forces similar to those provided by APL, without a significant increase in force structure. In certain situations, some alternatives might be useful: increased reconnaissance forward; more soldiers or weapons systems in a given battlefield area; more command-detonated Claymores to protect against a dismounted enemy; AT mines remotely delivered "just in time" to support a maneuver and inhibit the enemy's ability to breach; and speed, mobility, and offensive tactical operations.

Materiel Alternatives Against Dismounted Targets

Of all the current APL systems, only one is compliant with the Ottawa convention—the command-detonated Claymore munition. The Claymore scored well on both the military and the humanitarian criteria. However, the command-detonation feature limits the Claymore's ability to alert the force or provide early warning. The required set-up time makes it less effective than nonself-destructing APL for breaking contact with the enemy. On the positive side, from the humanitarian perspective, the Claymore may enable a soldier to discriminate between combatants and noncombatants before detonating the mine.

Materiel Alternatives Against Mounted Targets

Of the current mine systems designed for use against mounted targets, two are Ottawa complaint (RAAMS and Hornet/WAM), and one is Ottawa compliant when used in the pure AT mode (Volcano M87A1). At issue is whether any of them can perform as well without the APL that normally protect them.

The Volcano mine is produced in mixed and AT-pure versions. The mixed-system Volcano (M87) is the baseline against which alternatives for use against mounted targets were measured in this study. The AT-pure version (M87A1) is Ottawa compliant, but without APL, it is less effective. Although the AT pure Volcano is effective in destroying tanks and large vehicles, it does not have the militarily desirable characteristics of APL of slowing dismounted enemy breaches. The RAAMS system not accompanied by the ADAM would be less effective for the same reasons. Both the AT-pure Volcano (M87A1) and AT-pure RAAMS scored better than the baseline system (Volcano M87) according to the humanitarian criteria.

The Hornet/WAM is an AT-pure munition that is effective against its intended targets—heavy vehicles. Hornet/WAM is a novel weapon because it has a 100-meter lethal radius, which essentially negates an enemy's use of armored plow type breachers. Dismounted enemy forces have to clear a breach lane more than 200 meters wide in a Hornet/WAM minefield to allow armored vehicles to pass safely. The time for a dismounted enemy force to clear a lane this wide is likely to be similar to the time required for a dismounted enemy to breach a standard minefield of classic AT mines protected with APL. The Hornet/WAM is limited by the requirement that it be placed in position by hand and by its weight (15.88 kilograms) and high profile (35.6 centimeters).

TABLE 5-2 Score Sheet for Alternatives Available Today

		System Name	Claymore (M18)	Volcano (M87A1)	RAAMS	Hornet/WAM	Maverick	Longbow Hellfire	SFW	SADARM	BAT
		Available for Implementation	**N**	**N**	**N**	**N**	**N**	**N**	**N**	**N**	**N**
		Baseline D = M14/16 M = M87	D	M	M	M	M	M	M	M	M
	A0	**Military Effects** Sum:	**4**	**–2**	**–2**	**–3**	**–4**	**–4**	**–4**	**–4**	**–4**
Mounted	A1	Enhances effects of close and deep friendly fires		0	1	1	–4	–4	–4	–3	–3
Mounted	A2	Has multiple methods of delivery		0	–1	–1	–1	–1	–1	–1	–1
Mounted	A3	Provides a range of effects that inhibit mounted and dismounted maneuvers		–1	–1	–1	–2	–2	–2	–2	–2
Mounted	A4	Resists full spectrum of enemy breach methods, including dismounted methods		–2	–1	–1	n/a	n/a	n/a	n/a	n/a
Mounted	A5	Provides early warning of ground attack		0	–1	0	–1	–1	–1	–1	–1
Mounted	A6	Is safe for friendly forces		0	0	0	0	0	0	0	0
Mounted	A7	Is effective in all types of terrain and weather		0	0	0	–1	–1	–1	–1	–1
Mounted	A8	Poses minimal residual hazard to own forces and noncombatants after military conflicts		0	0	0	0	0	0	0	0
Mounted	A9	Is difficult to detect by enemy forces		0	0	–1	2	2	2	2	2
Mounted	A10	Poses minimal risk of fratricide		1	1	1	2	2	2	2	2
Mounted	A11	Has modifiable effects for mounted and/or dismounted threat		0	0	0	1	1	1	0	0
Mounted	A12	Has controllable activation/deactivation and duration before and after installation		0	0	0	1	1	1	1	1
Mounted	A13	Is effective in nuclear, chemical, and biological environments		0	0	0	–1	–1	–1	–1	–1
Mounted	A14	Is easy and efficient to distribute		0	0	–1	0	0	0	0	0
Dismounted	A15	Can delay, disrupt, and/or canalize enemy movement/maneuvers	1								
Dismounted	A16	Denies enemy access to terrain/facilities (including short- and long-term deterrent for boundaries and DMZ areas)	1								
Dismounted	A17	Enhances effects of friendly-force weapons, obstacles, and munitions (including AT mines)	1								
Dismounted	A18	Generates exploitable delays and opportunities (fixes or contains enemy)	1								
Dismounted	A19	Generates detection, alert, and/or early warnings	–1								
Dismounted	A20	Facilitates classification of target	1								
Dismounted	A21	Produces desired effects on enemy forces (nonlethal to lethal)	0								
Dismounted	A22	Reduces casualties/risk for U.S. and/or allied forces	1								
Dismounted	A23	Deters pursuit to facilitate breaking of contact under pressure	–1								
	B0	**Humanitarian Concerns** Sum:	**5**	**2**	**2**	**2**	**4**	**4**	**4**	**3**	**3**
	B1	Reduces postconflict residual hazard (CCW)	1	1	1	1	1	1	1	1	1
	B2	Location can be recorded (CCW)	1	0	0	0	0	0	0	0	0
	B3	Detectable to facilitate removal (CCW)	0	0	0	0	0	0	0	0	0
	B4	Discriminates between combatants and civilians (Ottawa)	2	0	0	0	2	2	2	1	1
	B5	Does not explode on presence, proximity, or contact of a person (Ottawa)	1	1	1	1	1	1	1	1	1
	C0	**Overall Technical Risk (C4ISR, etc.)**	**0**	**0**	**0**	**0**	**0**	**0**	**0**	**0**	**0**
	D0	**Requires Change in Tactics/Doctrine**	**N**	**N**	**N**	**N**	**N**	**N**	**N**	**N**	**N**
	E0	**Cost** Sum:	**0**	**1**	**1**	**–2**	**–1**	**–1**	**–2**	**–2**	**–2**
	E1	Research and development	0	0	0	0	0	0	0	0	0
	E2	Procurement	0	0	0	–2	–2	–2	–2	–2	–2

NOTE: Acronyms are defined on p. xv in the front matter.
Key: N = now, D = dismounted, M = mounted, n/a = nonapplicable.

The antihandling device provides some protection against tampering by a dismounted enemy. The Hornet/WAM fares quite well on the humanitarian criteria because it is self-destructing and self-deactivating and because it has no APL.

Five precision-guided weapons (Maverick, Longbow-Hellfire, SFW, SADARM, and BAT) were assessed as alternatives to mixed systems. All of them are capable of destroying armored vehicles and do not need the protection of APL. These air, missile, and artillery delivered weapons are well suited for their intended purpose of destroying massed enemy combat vehicles at significant distances from the close battle area. However, none would perform as well as the Volcano baseline. Their inability to distinguish between friendly and enemy armored vehicles precludes their use close to friendly troops, which limits their ability to enhance friendly-fire effects in close fight situations.

These weapons are all controlled at the brigade or higher level, which means there is likely to be a significant time delay between the identification of an enemy target and the arrival of the attacking munition. Precision munitions also have fairly limited kill zones. Therefore, to be effective after they are fired, the enemy vehicles must be in a relatively small elliptical kill area. Otherwise, the munitions fall to the ground and self-detonate. To prevent this, these weapons require highly accurate, real-time intelligence for targeting.

Other shortcomings of these weapons include the fact that they do not have multiple methods of delivery, have limited launch platforms, and do not provide a range of effects to inhibit dismounted maneuver. In addition, precision munitions delivered by air might negatively influence the larger battle space by forcing artillery and other indirect-fire systems to cease fire to ensure a safe air corridor for the delivery aircraft. Also, although not part of the scoring criteria, considerable concern was expressed about the unintended consequences of unexploded ordnance that might result from these weapons, the residual effects of which could be worse than the residual effects of the self-destructing and self-deactivating APL.

A system that was not scored by the committee but that deserves mention is the Air Force JSTARS (joint surveillance target attack radar system). The E-8C JSTARS is an airborne battle management and command and control platform that conducts ground surveillance to develop an understanding of the enemy situation and to support attack operations that contribute to the delay, disruption, and destruction of enemy forces. The primary mission of JSTARS is to provide dedicated support for high-level ground commander's (corps or division) intelligence requirements. The E-8C, a modified Boeing 707/300 series commercial airframe (the use of which is weather dependent), is a jam-resistant system capable of operating despite heavy electronic countermeasures. This system and its planned upgrades might provide a viable sensor alternative enabling small ground units to employ weapons to destroy enemy vehicles at a distance, thus reducing the need for certain types of landmines.

Summary

The criteria and scores are displayed in tabular form in Table 5-2. The details of how these scores were derived can be found in Chapter 4.

Figure 5-1 is a graphical summary of the scoring. In keeping with the Statement of Task, this graph shows only the relative military effectiveness of candidate systems without regard to cost, risk, or humanitarian factors. Each bar on the graph is a composite. The lower portion (white) shows the degree to which each system meets the military effectiveness requirements in comparison to the baseline

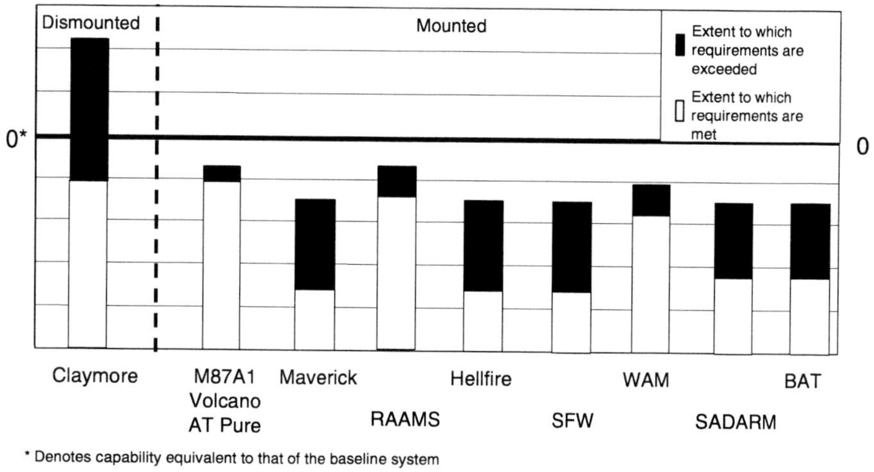

FIGURE 5-1 Military effectiveness of currently available alternatives based on qualitative scoring by the committee.

system. If the candidate system meets all of the requirements at least as well as the baseline system, the score is 0. If it is less effective in any requirement, the score is less than 0. The upper portion (dark shading) of the bar shows capabilities that exceed those of the baseline system.

For example, scores for the Claymore (in Table 5-2), show that 7 of 9 requirements were met or exceeded (i.e., had a score of 0 or higher); the lower (white) portion of the bar has a value of 7 out of 9, or 78 percent of requirements met. In addition, 6 of the 9 requirements were exceeded (i.e., had a score of 1 or more); the upper (black) portion of the bar has a value of 6 out of 9, or 67 percent.

A second example illustrates some additional factors. The scores for the Hellfire (again in Table 5-2) range from –4 to +2, which indicates significant deficiencies in some areas and significant advantages in others. To account for the impact of these larger excursions, the scores are adjusted by the degree of deficiency or improvement. Because a score of +1 is the lowest score that would be considered an "improvement," a score of +2 adds one additional point to the "improvements" score but does not affect the requirements score. Similarly, a score of –2 subtracts one point from the "requirements" score, and a score of –3 subtracts 2 points. In the case of Hellfire, 8 of 14 requirements are met (including the N/A as a "met" requirement), but deficiencies of 4 and 2 result in an adjusted requirements score of 4 out of 14, or 29 percent. Similarly, 4 of 14 requirements are exceeded, two by a score of +2, resulting in an adjusted improvements score of 6 out of 14, or 43 percent.

In general, if the total bar height in Figure 5-1 is high, the system is likely to be militarily effective. If the value of the lower portion of the bar is near 0, the system meets most of the military requirements. If the lower bar is much lower than 0, the system probably has significant differences from the baseline mine and will not perform some desired functions. However, that system may still be militarily effective if it performs some functions much better than the baseline system. Because the scoring criteria were not weighted, these graphs should be used only for assessing trends and making qualitative comparisons.

6

Alternatives Available by 2006

The Information Age will alter modern warfare in the twenty-first century just as the Industrial Age altered twentieth century battlefields with new forms of integrated mechanization. (Scales, 1999)

OVERVIEW

A variety of systems could be available for implementation by 2006. Some would be improvements of existing systems; some would be the results of DOD's early efforts to identify alternatives; and some would be combinations of systems that have been under consideration for other missions. However, unless DOD gives these new technologies a very high priority, six years will not be long enough for the weaponization of any innovative technology. These new alternatives will be characterized by the separation of sensors and kill mechanisms and improved communications between sensors and soldiers. In the near term, a "man-in-the-loop" will still be necessary, but sensors will alert him much earlier and more subtly than current APL, which only provide notification by exploding on contact. This chapter describes nonmateriel and materiel alternatives that could be available by 2006.

NONMATERIEL ALTERNATIVES

Some adaptations of tactics and doctrine may provide alternatives to certain aspects of APL, although none would replace all APL functions. In the near future, it may be possible to develop new tactics and doctrine that incorporate improvements in sensors and communications and provide more suitable alternatives than those currently available. However, the committee could not identify alternative tactics or operational concepts that could, on their own by 2006, provide tactical advantages similar to those provided by APL.

MATERIEL ALTERNATIVES

Common Features

Man-in-the-Loop

The man-in-the-loop feature requires that a soldier determine whether a target is friendly or hostile, visually or by using sensor input, before activating the response element. Although the man-in-the-loop satisfies humanitarian concerns and would reduce fratricide, it introduces some new vulnerabilities: (1) if the soldier/operator is disabled or killed, the minefield becomes inoperable; (2) the requisite communication links between the soldier/operator and the mines can be disrupted rendering the minefield ineffective; (3) a single soldier/operator may have difficulty processing large amounts of data under stressful, chaotic conditions; (4) some sensor capabilities, such as video imagery, are subject to degradation by adverse weather conditions or terrain; (5) sensors and communication systems could be susceptible to electromagnetic pulse.

Delinked Sensor-Shooter Concept

Separating the sensor from the response element (whether lethal or nonlethal) would provide a range of responses to an enemy breach. If the target can be identified prior to firing on it, the response can be more accurate and more effective. However, the more complicated the system, the more vulnerable it is to countermeasures. For example, some sensor capabilities, such as video imagery, are subject to degradation by certain adverse weather conditions or terrain. In addition, sensors and communication systems are more susceptible to electromagnetic pulse and similar phenomena than mechanical APL. Furthermore, systems in which the sensor is separated from the response element can not, by definition, respond instantaneously.

The committee evaluated 15 alternative systems that should be available by 2006. These are listed, along with some of their characteristics, in Table 6-1. Six of the alternatives would be used against dismounted threats. These were compared with the current M14 and M16 mines as a baseline. The nine alternatives that would be used against mounted threats were compared with the current M87 Volcano mixed mine system as a baseline. Descriptions of the alternatives follow as well as brief written assessments and a table measuring the alternatives against the criteria described in Chapter 4.

For Use Against Dismounted Threats

Hand-Emplaced Sensor Field

Source: Committee on Alternative Technologies to Replace Antipersonnel Landmines in discussions with DOD scientists

The Hand-Emplaced Sensor Field (HESF) could provide an early warning that dismounted intruders were approaching. The sensors could combine acoustic, laser, and infrared virtual trip wires, infrared imaging, motion, video, biological, seismic, or other technologies. Sensor input could be communicated back to the operator via either wire/fiber optics or wireless communication using a Single-Channel Ground-to-Air Radio System (SINCGARS).

The distance between the sensors and the friendly location would depend on the situation. Sensors would be far

TABLE 6-1 Alternatives Available by 2006

System Name	APL/ AT/ Mixed Non-Mine	Self-destructing/ Self-deactivating	Lethal/ Nonlethal	Ottawa Compliant[a]	Dismounted Enemy: Remotely Delivered	Dismounted Enemy: Hand Emplaced	Mounted Enemy: Remotely Delivered	Mounted Enemy: Hand Emplaced
Hand-Emplaced Sensor Field (HESF)	n/m	n/a	n/a	Y		X		
Nonself-Destructing Alternative (NSD-A) **Track I**	APL	Y	L	[b]		X		
Sphinx-Moder Perimeter Defense System	APL	N	L	Y		X		
Multiple-Shot Claymore Mine	APL	N	L	Y		X		
Bounding Nonlethal Munition (BNLM)	n/m	n/a	N/L	Y		X		
Taser Nonlethal Munition	n/m	n/a	N/L	Y		X		
Wide Area Munition Product Improvement Program (WAM PIP)	AT	Y	L	Y				X
Remote Area-Denial Artillery Munition (RADAM) **Track I**	Mix	Y	L	N	X		X	
RAAMS Enhanced with Telemetry (RD Telemetry)	AT	Y	L	Y			X	
Canister-Launched Area-Denial System (CLADS)[c]	APL	n/a	N/L	Y	X		X	
Volcano-CLADS	Mix	Y	L	Y	X		X	
AT Pure-Modular-Pack Mine System	AT	Y	L	Y			X	
AT Pure-Gator	AT	Y	L	Y			X	
Dual-Purpose Improved Conventional Munition (DPICM) with Random-Delay Fuzing (Popcorn)	Mix	Y	L	Y	X		X	
Small Short-Duration Mine System (SSDMS)	Mix	Y	L	N	X		X	

[a]The committee used the definition found in the Ottawa Convention to determine whether a system would be Ottawa compliant.
[b]Ottawa compliance would depend on whether the battlefield override switch was a part of the design.
[c]Assumed to be used alongside AT mines.

enough away to give the operator and his leadership time to assess the situation and take appropriate action. Responses could include alerting nearby outposts, dispatching a manned patrol, and sending a more sophisticated sensor array to gather more information. If enemy activity were confirmed, the operator would notify the leadership, who could take appropriate action, such as calling for direct or indirect fire.

HESF could reinforce the capabilities of the DOD Track I alternative (NSD-A) envisioned for use in Korea by giving soldiers time to react appropriately rather than being forced to take almost instantaneous action and having to be fully alert at all times.

With appropriate sponsorship, HESF with the following existing or emerging components could be ready by 2006: existing components of the Improved Remotely-Monitored Battlefield Sensor System, fiber optics, commercial anti-intrusion technology, and near state-of-the-art sensors that are being developed in governmental and developmental laboratories.

Doctrine and tactics for the HESF would have to be developed. Areas to be addressed would include: maximizing the use of terrain; reducing the enemy's ability to defeat the sensors; establishing patterns of sensor emplacement; and detailing how sensors would be integrated into the tactical plan to complement and take full advantage of other tactical weapons in the battlefield area.

The most difficult challenge is likely to be in hardening and integrating the various components. The Army has already conducted some experiments on the HESF concept, and the research community has invested heavily in sensor technologies in the past several years. However, there is currently no funded program for developing an HESF.

Advantages

- HESF would be compliant with the CCW Amended Protocol II and the Ottawa Convention.
- HESF would provide early warning, allowing enough time for a coordinated tactical response that could bring in all available weapon systems.
- The system has a man-in-the-loop, which may address humanitarian concerns.
- The additional reaction time may reduce fratricide and injuries to noncombatants.

Disadvantages

- Hand emplacement is time consuming.
- HESF would require that new tactics and doctrine be developed.
- Differentiating among enemy soldiers, noncombatants, friendly forces, and animals could be difficult.
- HESF would have higher research, development, and procurement costs than other alternatives in this category.
- No resources for development and production are currently available.
- HESF could not be used for deep operations because sensors would have to be close to friendly forces.
- HESF would have to be supported with other weapons systems.

The Track I Alternative to Nonself-Destructing APL

Source: U.S. Department of Defense/Office of the Project Manager, Mines, Countermine, and Demolitions, briefings to Committee on Alternative Technologies to Replace Antipersonnel Landmines (Rosamilia, 2000)

The DOD Track I initiative has begun the development of an alternative (NSD-A) to nonself-destructing APL, which uses co-located sensors and kill mechanisms controlled by a man-in-the-loop. When activated by an intruder, the sensor sends a signal to the soldier/operator's control panel (a repeater may be necessary to relay the signal). Depending on the rules of engagement, an observer may oversee the area to ensure positive identification. The kill mechanism is a munition, much the same size and with much the same effect as the current M14/M16, but may be augmented or replaced by a nonlethal weapon.

NSD-A systems are being developed to meet the requirements of the CINC in Korea for protective APL minefields. One option requested by the CINC is the ability to put the mines in an autonomous mode (i.e., to remove the man-in-the-loop) in certain tactical situations, such as when the commander anticipates an attack by an overwhelming enemy force. Technically, this can be accomplished by altering the software used in the system. Engagement of this so-called "battlefield override switch" would render the system noncompliant with the Ottawa Convention (see Box 6-1 on p. 55). If the switch was engaged, the NSD-A would self-destruct in four hours.

Doctrine and tactics would have to be revised to address the NSD-A alternative, and all of the proposed NSD-A systems would require extensive research, development, testing, and evaluation programs, which are fully funded at present.

Advantages

- NSD-A would comply with the CCW Amended Protocol II.
- Without the battlefield override switch, NSD-A would comply with the Ottawa Convention.
- The system has a man-in-the-loop, which should minimize friendly and noncombatant casualties.
- NSD-A could include nonlethal options that may alleviate humanitarian concerns and allow for more military options.
- The self-destruct feature alleviates humanitarian concerns.
- The early warning provided by the sensor would be at least equivalent to warnings by most APL.

Disadvantages

- If the system includes the battlefield override switch, it would not comply with the Ottawa Convention.
- The time required for system emplacement, as well as the reaction time of the man-in-the-loop, are unknown.
- Because mines must be hand emplaced, the system would not be practical for use beyond immediate defensive positions.

Sphinx-Moder Perimeter Defense System

Source: Data from French Embassy (Etienne Lacroix Defense, 1998)

This French system is described as a way to improve the protection of fixed assets, troop concentrations, and command centers. The control and firing panel, with power supply, is connected to the firing unit by a 150-meter connecting cable. Three types of ammunition can be used: fragmentation rounds, warning (flash-bang) rounds, and training rounds. The firing unit is loaded with three canisters, each containing two fragmentation or warning "spheres," which are launched in a 140-degree arc a distance of 50 meters (fragmentation) or 80 meters (warning). The fragmentation spheres appear to detonate at ground level indicating either a time-delay or impact fuze. The warning round is designed to produce an air burst. A number of devices can be combined into a comprehensive system with a central control panel to provide interlocking coverage.

The Sphinx-Moder appears to be a ground adaptation of a vehicle-mounted system. No sensors are described in the product literature, so the identification of targets requires a man-in the-loop as an observer. It appears that the fire command will fire all three dispensers on the launcher, which can be reloaded, and an unfired unit can be relocated.

In several respects, the principles of operation appear to be similar to those of the Claymore. The Sphinx is a man-portable unit (20 kilograms, 240 millimeters high, 550 millimeters long, 480 millimeters wide) that can be operated on any type of surface, set up for operation within 10 minutes, and operated in an urban area or any other area without cover. U.S. doctrine and tactics would have to be revised to address this alternative.

French industry is marketing the Sphinx as an available system, and photographs in the literature indicate that at least one prototype has been produced. It is not known if additional research, development, testing, and evaluation would be required before fielding. The ammunition appears to be adapted from a vehicle application and, therefore, should be available.

Advantages

- Based on the limited information available, the Sphinx appears to comply with CCW Amended Protocol II and the Ottawa Convention.
- The Sphinx has a man-in-the-loop, which should minimize friendly and noncombatant casualties.
- The presence of the Sphinx, which can disarm and/or maim an enemy, should have a psychological impact, although probably less of an impact than APL.

Disadvantages

- Sphinx would not provide an early warning of enemy presence.
- Sphinx requires a soldier or a separate sensor package to provide a warning equivalent to the warnings provided by most APL/APL alternatives.
- The time required for system emplacement, as well as the reaction time of the man-in-the-loop, are unknown.
- Sphinx would not be appropriate for use in tactical minefields because the mines must be hand emplaced, which would be difficult beyond immediate defensive positions.

Multiple-Shot Claymore Mine

Source: Committee on Alternative Technologies to Replace Antipersonnel Landmines

The present Claymore mine has two disadvantages: it is a single-shot munition, and its cone of fire is fixed upon emplacement. The multiple-shot Claymore mine would be a three-shot version that could be modified in elevation and azimuth by a small, battery-powered electric motor. The multiple-shot Claymore mine is dependent on a rocket-propelled projectile rather than an explosive-backed projectile. The rocket grain would have to be of sufficient size for the projectile to achieve a velocity adequate to penetrate body armor. Three Claymores would be positioned back-to-face in a light aluminum or plastic frame. The set-back of the rocket exhausts would release a spring that would kick the spent mine out of the frame and move the next mine into firing position.

The multiple-shot assembly would be equipped with an infrared sensor (plus optical sensor used in daylight) for surveillance of the field of fire. To conserve energy, a forward placed trip wire would alert the surveillance sensor. The surveillance scene would be transmitted to a remote observer who would determine if the breach was by friend or foe and would give the order to fire for each round, as appropriate.

The deployment of the multiple-shot Claymore mine would be the same as for the Claymore, but new tactics and doctrine would have to be developed. All technology is within the state of the art, although some research and development on the rocket-propelled projectiles would be necessary. The multiple-shot Claymore might be available by 2003.

Advantages

- The system does not have the single-shot, fixed field of fire limitation of the current Claymore.

- The multiple-shot Claymore would comply with the CCW Amended Protocol II and the Ottawa Convention.
- Human operators would discriminate between friend and foe.
- Mines could be neutralized by removal for reuse.

Disadvantages

- Together these mines would be heavier and bulkier than a single-shot Claymore, which would make it more difficult for a dismounted soldier to carry it into battle and conceal.
- The multiple-shot Claymore has higher research, development, and procurement costs than other alternatives in this category.

Bounding Nonlethal Munition

Source: Existing Developmental System (Irish, 2000)

The Bounding Nonlethal Munition (BNLM), a joint Army/Marine Corps nonlethal project, is a hand-emplaced, low-hazard, low-shrapnel-producing, target-initiated munition. The system is designed to confuse and disorient an enemy force. BNLM is a recoverable, reusable, aluminum-cased munition that comes in three different types designed to deliver three separate payloads—rubber balls, a kevlar net, and paint balls. BNLM requires a Canister-Launched Area-Denial System (CLADS) launcher.

The BNLM is designed for use in military operations other than war such as countering civil disturbances, protecting the site and area around key facilities, and dealing with threats before a situation escalates and requires lethal force. A limited number of BLNM are included in the Army Nonlethal Capability Set and the Marine Corps Capability Set for contingency operations.

The BNLM is in the concept exploration phase of development. If munitions with different payloads can be standardized, all of them could be delivered via the CLADS launcher.

Advantages

- Even though BNLM explode on contact, presence, or proximity of a person, they should still be acceptable under the Ottawa Convention because their payloads are nonlethal.
- BNLM complies with the CCW Amended Protocol II.
- Maintenance and storage costs are expected to be moderate.
- Nonlethal payloads will alleviate humanitarian concerns and reduce friendly casualties.

Disadvantages

- The effects of nonlethal devices on humans are not fully known.
- Because BNLM is not lethal, it will have a limited psychological or physical effect on a determined enemy.
- Pellets cannot penetrate a shield or body armor.

Taser Nonlethal Munition

Source: Briefings by the Department of Defense/Office of the Project Manager, Mines, Countermine, and Demolitions (Persau, 2000)

The conceptual Taser munition would use the same nonlethal "stun" technology currently used by civilian police to temporarily incapacitate uncooperative subjects. Upon activation, an audible alert would indicate the presence of an intruder. The sensor would automatically identify the intruder whether standing or prone, day or night, at a range of up to 5 to 6 meters in its sector of surveillance. Once triggered, the Taser would fire a pair of electrically charged darts propelled by a rifle round primer. The darts would be connected to a battery and other electronic circuitry in the munition by thin 6-meter wires. Once the two pronged darts attach to skin or clothing, they would deliver an incapacitating electric shock of 50,000 volts in 4 to 6 microsecond pulses 10 to 20 times per second.

The current power supply can support approximately 10 minutes of continuous operation. However, because the desired effect is achieved in a few seconds, and because recovery from the Taser is not immediate, the electric shocks could be cycled to extend the time of incapacitation to several hours. With low-power-consumption technologies, the Taser could operate for extended periods of time. Vital components could be configured with a self-destructing feature. Testing by the U.S. Army Tank-Automotive and Armaments Command's Armament Research, Development, and Engineering Center has shown that even if the dart is affixed to military clothing, heavy body armor, or a kevlar helmet, the electrical shock will still be conducted to the target. The hand-emplaced version of the Taser, which would be lightweight and pocket-sized, would have an on-off capability so it could be redeployed. Dispensed Tasers are not expected to be recoverable.

The Taser munition could also be deployed via a launcher currently used for existing scatterable mines. The hand-emplaced munition would house one Taser unit triggered by trip wire; the launched version would house two Taser units oriented 90 degrees from each other, operating independently, and triggered by a passive infrared sensor. Tasers could provide nonlethal, APL-like protection for key positions, facilities, or AT mines.

Many of the components required to build the Taser already exist. Depending on user requirements, a moderate developmental effort could be required to move the munition into the production phase. Current research and development are being done as part of the Fiscal Year 1999 Joint

Non-Lethal Weapons Directorate Technology investment proposal led by the Armament Research, Development, and Engineering Center. The contractor has demonstrated an initial concept prototype. No funded follow-on effort is planned, and no user Operational Requirements Document (ORD) for a formal development or acquisition program has been developed.

Advantages

- Because electric shocks are considered to be nonlethal, Tasers should be acceptable under the Ottawa Convention.
- Tasers should also be compliant with the CCW Amended Protocol II.
- Little or no collateral damage is associated with the Taser.
- Maintenance and storage costs are expected to be moderate.
- Existing platforms now used to deliver scatterable mines could be used to deliver Tasers.
- The hand-emplaced version can be turned off and redeployed.
- Nonlethal effects would alleviate humanitarian concerns and minimize friendly casualties.

Disadvantages

- The effects of nonlethal devices on humans (particularly children) are not fully known. However, the experience of police forces is a rich source of data that may not be available for other nonlethal technologies.
- Taser munitions are not lethal and, therefore, will have a limited psychological or physical effect on a determined enemy.
- Intruders could deflect the darts with antiriot shields.
- Hand emplacement is time consuming and manpower intensive.
- The dispensed version is not redeployable.

For Use Against Mounted Threats

Hornet/Wide Area Munition Product Improvement Program

Source: Briefings by the U.S. Department of Defense/Office of the Project Manager, Mines, Countermine, and Demolitions (Strano, 2000)

The Hornet/Wide Area Munition product improvement program (WAM PIP) is a two-phased evolutionary program to improve the Hornet/WAM hand-emplaced AT mine. Improvements include an on/off capability, man-in-the-loop overwatch, and improved sensors. The first phase, the A1 version, would place a man-in-the-loop. The A1 would use the line-of-sight SINCGARS to connect a series of Hornet/WAMs to an operator-controlled, hand-held, terminal-unit laptop computer. With the embedded Global Positioning System (GPS) capability, the operator would be able to view an electronic map of the field showing the exact location and status of each mine. The Hornet/WAM minefield and the hand-held terminal unit could be separated by three to five kilometers. The operator would be able to turn the Hornet/WAM on remotely, then off, then on again; while off, friendly forces or noncombatants could cross a minefield safely. In the on mode, the mines would operate autonomously.

The operator would also be able to activate the sensors remotely as sentries or arm, disarm, redeploy, or self-destruct each munition. In the sentry mode, the sensors would signal the hand-held terminal unit when heavy or light vehicles were detected, but the munition would not attack unless commanded to do so. Each Hornet/WAM could be redeployed prior to initial arming. Current self-destructing and antihandling capabilities would be retained in the A1 version, which is currently scheduled to be type classified as standard in the second quarter of Fiscal Year 2002.

The second phase of the WAM PIP, the A2 version, would have an improved sensor to improve target detection and an improved dual-purpose warhead that would perform better against "soft" targets, as well as heavy armor. Both of these improvements are being developed by the Air Force SFW PIP. The A2 is currently scheduled to be type classified as standard in the fourth quarter of Fiscal Year 2003. The A2 version (like the A1 version) would require hand emplacement.

The Hornet/WAM is currently used against armored forces in protective and tactical minefields. As improved versions of the munition are fielded, changes would have to be made to doctrinal publications. The WAM PIP is a comprehensive, fully funded, research, development, testing, and evaluation program.

Advantages

- The WAM PIP would comply with the CCW Amended Protocol II and the Ottawa Convention.
- The system has a man-in-the-loop, which should reduce friendly and noncombatant casualties.
- The WAM PIP has a sensor that would provide detection information to the operator.
- The WAM PIP would have an antihandling device to inhibit tampering.
- The self-destructing feature and command on-off should address humanitarian concerns.

Disadvantages

- Hand emplacement would severely limit the utility and range of the WAM PIP on the battlefield.
- The projected unit cost is high compared to the cost of standard AT mines.
- Line-of-sight communications would limit the use of the WAM PIP.

Remote Area-Denial Artillery Munition (RADAM)

Source: Briefing by the U.S. Department of Defense, BRTRC Technical Research Corporation (Bornhoft, 1999)

The proposed Remote Area-Denial Artillery Munition (RADAM) would combine the existing RAAMS (AT mines) and ADAM (APL) into one 155-mm howitzer projectile. The rationale for this combination is to provide AT mines with the same protection APL provides against dismounted breaching attempts and adhere to the presidential policy of not using pure APL outside of Korea after 2003. Existing doctrine and tactics should be adequate for RADAM, although adjustments to publications and firing tables would have to be made. RADAM would require research and development to adapt existing systems and an extensive redesign testing program.

Advantages

- RADAM would be an artillery-deliverable mixed system compliant with current presidential policy.
- RADAM would comply with the CCW Amended Protocol II.
- The combined AT / APL projectile would have efficient single-tube delivery and simplified logistics requirements.

Disadvantages

- RADAM would not comply with the Ottawa Convention.
- A significant number of artillery assets would be required to emplace a large minefield.
- RADAM would have no command-destruct feature, and the self-destruct times could not be extended by command.

Remote Antiarmor Mine System Enhanced with Telemetry

Source: Committee on Alternative Technologies to Replace Antipersonnel Landmines

Remote Antiarmor Mine System (RAAMS) enhanced with telemetry (RD-Telemetry) would add a subminiature telemetry and communications package to the existing RAAMS system. The telemetry would calculate the precise location of the dispensed mines (e.g., out to approximately 20 kilometers from a firing howitzer and hundreds of meters above ground) by using the GPS, inertial guidance, or other methodologies. The data communicated back to friendly forces by a miniature radio inside the projectile would provide a precise estimate of the ground location of scattered mines. The information could also be transferred directly to combat digital command and control systems to update digital maps in the battlefield sector automatically with symbols showing the locations and self-destruct times of the mines.

RD-Telemetry would have two principal benefits: (1) it would increase friendly situational awareness and thereby reduce fratricide caused by friendly soldiers unknowingly entering one of their own AT minefields; (2) after hostilities, more accurate locations could be used to plot mined areas and confirm that all mines have self-destructed.

This technology could be applied as a proof of principle in the RAAMS and then applied to all mine payloads, including mines dispensed by ground vehicles (Volcano), mines dispensed by aircraft (Gator and Volcano), and other submunitions. More complex communications would then be necessary to relay data from on-the-ground mines to a data-fusion or command center.

The system would be delivered by existing 155-mm howitzer projectiles and other systems used to remotely deliver AT mines. With near real-time knowledge of the location of emplaced minefields, remotely delivered AT minefields could be used "just in time" to support mobile forces and other operations. This would reduce the time an enemy has to locate and clear artillery-delivered AT mines and reduce the need for integrated APL to protect the AT mines. Antihandling devices on 20 percent of the munitions would discourage tampering.

Upgrading the RAAMS projectile and developing the capability of receiving and processing signals from the telemetry device are likely to require at least moderate research and development. Although most of the components to develop this system exist, integration with existing systems will require careful planning. Integrating transmitted data into existing command and control digital systems would also require moderate research and development.

Advantages

- RD-Telemetry would comply with the CCW Amended Protocol II and the Ottawa Convention.
- Twenty percent of these mines would have an antihandling device to inhibit tampering.
- The self-destruct feature would help alleviate humanitarian concerns.
- A more precise estimate of the ground location of scattered mines could reduce fratricide caused by friendly forces inadvertently entering a friendly minefield.
- A more precise estimate of the ground location of remotely delivered mines could be helpful for confirming that mines had (or had not) self-destructed.
- If combat commanders were confident of the location of scattered mines in a critical area, they might be able to use remotely delivered AT mines "just in time" thus decreasing the chance that the enemy could find and clear the minefield.

Disadvantages

- A significant number of artillery assets would be necessary to emplace a large minefield.

- The system would not have a command-destruct feature, and self-destruct times could not be extended by command option.
- Research, development, and acquisition costs are high.
- No existing research and development program could lead to the technology required for this enhancement.

Canister-Launched Area-Denial System (CLADS)

Source: Briefings by the U.S. Department of Defense/Office of the Project Manager, Mines, Countermine, and Demolitions (Irish, 2000)

Canister-Launched Area-Denial System (CLADS), a joint Army/Marine Corps nonlethal program, will be based on the Volcano multiple-delivery mine system. The launcher rack will hold 20 canisters, each of which contains four or five nonlethal munitions that can be used at extended stand-off ranges. These nonshrapnel-producing munitions will deploy trip wires upon landing. When activated, they will emit an audible warning and fire .32-caliber nonlethal rubber balls. CLADS will have a 360 degree radius of 5 to 15 meters and will be delivered by High-Mobility Multipurpose Wheeled Vehicles (HMMWVs) or rotary-wing aircraft. Production versions are envisioned to self-destruct approximately 48 hours after deployment.

CLADS is expected to be used in upcoming warfighting experiments. Plans to use this system include military operations other than war, civil disturbances, protection of key sites and facilities, and dealing with threats before a situation escalates to the point that lethal force must be used. Although CLADS is in the concepts exploration phase of research and development, it is currently on hold.

Advantages

- Even though CLADS will be activated by contact, presence, or proximity of a person, they should be acceptable under the Ottawa Convention because their payloads are nonlethal.
- CLADS complies with the CCW Amended Protocol II.
- The use of nonlethal payloads will alleviate humanitarian concerns and reduce friendly casualties.
- Employment will be relatively rapid.
- The presence of a self-destruct feature will alleviate humanitarian concerns.

Disadvantages

- The effects of nonlethal devices on humans are not fully known.
- Because CLADS is nonlethal at normal engagement ranges, its psychological and physical impacts will be limited.
- Research, development, and procurement costs will be high.

Volcano-CLADS

Source: Committee on Alternative Technologies to Replace Antipersonnel Landmines

The committee proposes that the Volcano system be modified to combine one CLADS munition (in place of the APL) and five standard Volcano AT mines in one canister. The Volcano-CLADS system would be used in protective and tactical minefields in all engagement areas on the battlefield. The mines would be laid out in a random distribution and would self-destruct at a selected time. The AT mines use a magnetic-influence fuze and a self-forging fragment to provide full-width coverage beneath armored targets. The nonshrapnel-producing CLADS munitions would deploy trip wires upon landing and would emit an audible warning and .32-caliber nonlethal rubber balls, when activated. Self-destruct times would be set at launch for 4 hours, 48 hours, or 15 days. The system would cover a 360-degree radius of 5 to 15 meters.

The dispensers could be mounted on several tracked or wheeled vehicles or on UH-60 Blackhawk helicopters. The system would consist of the launcher rack and dispenser control unit, vehicle-specific mounting hardware, and mine canisters. A completely loaded dispenser would hold 160 canisters, or 960 mines. The mines would be placed in a uniform density, approximately one mine per linear meter over an area of one square kilometer. The minefields could be emplaced anywhere on the battlefield reachable by the dispensing vehicles.

The Volcano-CLADS would not be covered in current doctrinal manuals; the development of tactical employment options would also be necessary. Development of the CLADS munition would have to be completed and integrated into the modified Volcano canister. The Volcano launcher and the Volcano AT mines are already developed.

Advantages

- Both the AT mines and the nonlethal CLADS would comply with CCW Amended Protocol II and the Ottawa Convention.
- The presence of a self-destruct feature would help alleviate humanitarian concerns.
- The nonlethal CLADS would alleviate humanitarian concerns and minimize friendly casualties, although the AT mines would be lethal.

Disadvantages

- The absence of a lethal APL component could reduce the time it would take a dismounted force to breach the AT minefield.
- The effects of nonlethal devices on humans are not fully known.
- The CLADS would not be lethal at normal engagement

ranges and would, therefore, have a limited psychological or physical impact.
- Research, development, and procurement costs would be high.

Antitank-Pure Modular-Pack Mine System

Source: Committee on Alternative Technologies to Replace Antipersonnel Landmines

The Modular-Pack Mine System (MOPMS) is a man-portable, 73.6 kg "suitcase" dispenser, normally used to close gaps in large minefields or to establish point minefields. It is technologically feasible to remove the APL from a standard MOPMS to create this variant, although more MOPMS would be required to create a large tactical minefield than are normally available to a unit. Each man-portable dispenser in the AT-pure version would contain 21 AT mines that could be dispensed on command by a signal sent either by wire or from a special radio transmitter (M71 remote control unit). If the mines were not dispensed, the dispenser could be relocated and used in another tactical situation. The mines would have a set self-destruct time of four hours; the self-destruct time could be extended three times. The mines could be command detonated if the tactical situation dictated.

Tactical use of the current MOPMS is covered by current doctrine; some modifications might be necessary to cover AT-pure MOPMS. The removal of APL and the reconfiguration of MOPMS to an AT-pure configuration should not require major research and development. However, modernization of the communications in the launcher and the M71 remote control unit would require research and development.

Advantages

- AT-pure MOPMS would comply with the CCW Amended Protocol II and the Ottawa Convention.
- The presence of a self-destruct feature would address humanitarian concerns.

Disadvantages

- The small number of mines in the total inventory could limit the overall utility of this system.
- The absence of APL is likely to reduce the time required for a dismounted force to breach a MOPMS minefield.

Antitank-Pure Gator

Source: Committee on Alternative Technologies to Replace Antipersonnel Landmines

The current Gator, which is designed for use in tactical minefields in all engagement areas on the battlefield, can be used wherever a tactical aircraft can reach. It is technologically feasible to remove the APL from the current mixed Gator system to create an AT-pure variant. The AT mine uses a magnetic-influence fuze and a self-forging fragment to provide full-width coverage beneath armored targets. Air Force and Navy versions differ only in the number of mines in the tactical munition dispensers used to deliver the mines.[1] Each mine has a square-shaped aeroballistic protective casing designed to aid dispersion. Self-destruct times, set by the pilot prior to release, are 4 hours, 48 hours, or 15 days. Although Gator can be used in close combat situations, its intended use is for deep or interdiction targets.

Tactical use of the current system is covered in current doctrinal manuals; some modifications may be necessary for an AT-pure system. The removal of the APL and the reconfiguration to an AT-pure system should not require major research and development.

Advantages

- The AT-pure Gator system would comply with the CCW Amended Protocol II and the Ottawa Convention.
- The system would have an antihandling device to deter tampering; this device could kill or wound an enemy.
- The self-destruct feature would address humanitarian concerns.

Disadvantages

- The absence of APL is likely to decrease the time required for a dismounted force to breach the minefield.

Dual-Purpose Improved Conventional Munitions with Random-Delay Fuzing

Source: U.S. Department of Defense scientists, Vietnam Veterans of America Foundation

The Dual-Purpose Improved Conventional Munitions (DPICM) with random-delay fuzing (sometimes called "popcorn") was suggested as an alternative by the Vietnam Veterans of America Foundation (Rossiter, 1999). Each remotely deployed DPICM submunition would be fitted with a delay fuze set to detonate randomly after dispersal from the carrier (hence the popcorn effect). Submunitions would be dispersed over already remotely delivered AT-pure minefields containing RAAMS, Gator, or Volcano.

Random detonations of the submunitions would deter dismounted breaching attempts during the random-delay period, which could be set to coincide with the self-destruct time of the AT mines. Casualties would be caused by controlled fragmentation of the case. If an enemy force was

[1] The Air Force dispenser, the CBU-89/B, contains a total of 94 AT mines, whereas the U.S. Navy dispenser, the CBU-78/B, contains 60 AT mines.

known to be enroute to the area, the delay fuze could be set for a shorter time, such as 30 minutes.

Current DPICM can be delivered by the MLRS, which would deliver 12 rockets, each containing 644 DPICM or by two 155-mm artillery projectiles, an M483 containing 88 DPICM or an M864 containing 72 DPICM.

Doctrine and tactics would have to be revised for the popcorn version of the DPICM, and the development of the random-delay fuze would require a research, development, testing, and evaluation program.

Advantages

- DPICM/popcorn would comply with the CCW Amended Protocol II and the Ottawa Convention.
- Because of its random activation, this system would have a psychological impact on an enemy.
- The random detonation could disrupt normal dismounted breaching methods (e.g., probing, hand grapnel, rifle-fired grapnel, or line charge).
- DPICM/popcorn could be used alone as an area-denial weapon.

Disadvantages

- DPICM/popcorn would require two launch platforms if combined with AT mines delivered by Gator or Volcano.
- DPICM/popcorn would be difficult to use close to friendly troops because the delivery system is imprecise.
- DPICM/popcorn would have a very small lethal radius and would, therefore, require a very large expenditure of ordnance.
- High-density distribution may be necessary to achieve the desired effect, which could entail diverting artillery assets from higher priority targets.
- If the MLRS were used for delivery, a minimum of six rockets would have to be loaded with DPICM, thus reducing the flexibility of the MLRS.
- Because the submunition would be exploded while on its side (not the designed position), it would be less effective than a normally deployed DPICM.
- Because the submunition detonates randomly, it is not as lethal as APL.
- Random detonation of popcorn submunitions could cause adjacent AT mines to detonate.
- Random detonations could alert the enemy to the location of the minefield.

Small Short-Duration Mine System

Source: Briefing by U.S. Department of State, Arms Control and Nonproliferation Advisory Board (Garwin and Sherman, 2000)

The Small Short-Duration Mine System (SSDMS) would be a hand-emplaced canister containing a combination of current scatterable APL and AT mines similar or identical to Gator/Volcano mines. The mines would remain in the canister until a secure radio command was received, at which time they would be pyrotechnically ejected, emplaced, and activated. From this point forward, the mines would be activated by the target. If the mines were not triggered within a preset period of time (30 days maximum after emplacement), they would self-destruct by exploding. If for some reason the self-destruct mechanism failed, they would self-deactivate when the batteries ran down, in 120 days or less.

The SSDMS would consist of three units: the canister, the mines, and a radio receiver. The canister, about 12 centimeters in diameter, 5 to 6 centimeters high, and weighing about 16 kilograms, would house six mines and the radio receiver. The canister is essentially a single Volcano tube with a radio receiver added. With intermittent duty cycles, canister battery life could be extended to several years. The overwatch control unit (the radio) would allow a soldier/operator to emplace the mines by command. Variants of the basic design could be optimized for steeply inclined surfaces.

To create a denied zone, the canisters would be installed upon warning of an imminent attack, as the United States now plans to do with its hand-emplaced, nonself-destructing APL in Korea. Installation of SSDMS would be more rapid than with present mines (because six mines could be installed at once) but would require more digging. SSDMS could also be installed in peacetime because, unlike M14 and M16 APL, the SSDMS canister would be harmless until further action is taken. Because the mines themselves would not be emplaced, the canisters could remain in their installed positions indefinitely under the terms of the CCW Amended Protocol II. The security of the canisters could be improved by burying them, which could probably be done rapidly using a post-hole digging device.

When required, the mines would be emplaced by firing the canisters via a remote command from a secure radio signal. The commander would have the option, which he does not have now with conventional mines, of laying all of the mines in a field at once or emplacing some and holding the rest in reserve for emplacement under fire without risking his own troops.

Unlike today's persistent mines, SSDMS mines would present no long-term hazard. When the threat had receded or the conflict had ended, unfired canisters could either be left in place for future contingencies or recovered for later use. For SSDMS to be effective, communications would have to be secure.

Advantages

- SSDMS would be an all-weather, all-terrain system that used familiar, existing components.
- SSDMS would comply with the CCW Amended Protocol II.

- Technical risk and costs would be very low.
- Development would require two to three years, most of which would be spent in testing.
- SSDMS would enable the United States to retire its entire stock of persistent APL and, perhaps, some of its persistent AT mines as well.

Disadvantages

- Although SSDMS mines could be remotely emplaced, the canisters would have to be hand emplaced.
- SSDMS would not comply with the Ottawa Convention.
- Covert emplacement would be time consuming.

COMMITTEE ASSESSMENTS

Materiel Alternatives Against Dismounted Targets

Lethal Alternatives

Three lethal alternatives could be available for use against dismounted forces: the multiple-shot Claymore, the French Sphinx-Moder perimeter defense system, and DOD's Track I initiative, NSD-A. The first two scored relatively well; the third is discussed in considerable detail in Box 6-1. Both the multiple-shot Claymore and the Sphinx-Moder would be improvements over the current Claymore. For area-denial, either would be more effective than the M14 and M16; however, neither would provide any early warning.

The Track I alternative NSD-A could provide, by 2006, similar or enhanced tactical advantages for U.S. forces as compared to those provided by current M14 and M16 mines. The battlefield override switch, a software capability that allows the system to operate autonomously, is highly contentious because, as presently designed, it would render the NSD-A non-Ottawa compliant. The issues surrounding this switch are explained in some detail in Box 6-1. Even though the timing of a decision on the switch or other programmatic delays could jeopardize the timeline, the NSD-A system appears to be technically mature enough to be available by 2006. This weapon system could be greatly enhanced in the future by planning for the inclusion of additional sensors, nonlethal elements, and an Ottawa-compliant battlefield override capability.

BOX 6-1
Track I Nonself-destructing Alternative (NSD-A)

The NSD-A was scheduled to enter the engineering and manufacturing development phase in the fall of 2000. However, the design of the NSD-A system has not been finalized pending a decision on whether or not to include the "battlefield override switch" capability. The battlefield override switch would permit the soldier/operator to activate software that would place the system in an autonomous mode. With this feature engaged, the man-in-the-loop would no longer be required to activate the munition, which would become a conventional, target-activated APL. Because this is a very contentious issue, DOD is conducting a separate study as part of its decision-making process. As of January 2001, this study had not been completed.

The purpose of the battlefield override switch is to prevent U.S and friendly casualties. However the inclusion of the switch would render the NSD-A system non-Ottawa compliant. Several rationales have been proposed for including it: (1) in some situations an extremely large enemy force might present so many targets that the NSD-A soldier/operator would be overwhelmed with information and unable to operate the system effectively; (2) in the event of a withdrawal under fire, placing the mines in an autonomous mode could extend their utility by inflicting casualties and delaying the enemy without requiring that the soldier/operator remain in the area; (3) the autonomous system would protect unobservable AT mines used to bolster defenses along lightly defended, economy-of-force sectors.

The committee was informed that either version of the NSD-A software is technologically feasible. Therefore, the issue is really political, based on how strongly the United States is committed to complying with the Ottawa Convention. The NSD-A with the battlefield override switch would provide the military with greater flexibility in responding to enemy threats. However, the Ottawa-compliant version of the NSD-A without the battlefield override switch appears to have significant tactical advantages over existing M14/M16 APL and would reduce the potential for fratricide and noncombatant casualties.

Recommendation. The development and production of the Track I alternative to nonself-destructing landmines (NSD-A) system should be aggressively pursued to ensure its availability by 2006.

Recommendation. Two suites of weapon software should be developed simultaneously in preparation for a presidential decision concerning the Ottawa Convention. If compliance with the Ottawa Convention were desired, the battlefield override switch, as currently designed, would not be used in the production of the NSD-A. If the president decides that other considerations outweigh Ottawa compliance, the option of retaining the switch would be available. In any case, Ottawa-compliant variations to the battlefield override switch should be explored to provide the United States with greater flexibility.

Nonlethal Alternatives

As discussed in Chapter 3, nonlethal variants by themselves cannot replace APL. Although a Joint Staff-level directorate has been established to explore the feasibility of nonlethal weapons, few resources have been allocated for this purpose. Nonlethal alternatives assessed by the committee scored lower than APL because they are not likely to arouse the life-threatening terror characteristic of lethal munitions or cause the serious disruptions brought about by casualties. In general, nonlethal weapons are less likely to deter a determined enemy, although in certain military operations, nonlethal variants may be useful.

Two promising nonlethal alternatives, BNLM and the Taser nonlethal munition, were considered as hand-emplaced deterrents to dismounted threats. Both weapons, which could be developed as remotely delivered devices, would provide protection against dismounted breaches of AT minefields by warning of attempted breaches and slowing the enemy long enough to enable U.S. forces to complete a maneuver or bring in additional fire. A remote deployment capability for the BNLM is currently under consideration.

Sensor System

HESF could alert a soldier/operator of an intruder; the operator could then monitor the situation via the sensors and activate the kill mechanism when and if necessary. Whereas APL have a very limited radius of effects, the HESF would bring available combined arms to bear much earlier.

HESF illustrates the significant added value of sensor systems closely integrated with a communications package and a kill mechanism for various APL alternatives. Some form of kill mechanism would have to be used with this system to make it complete. As sensor technologies are developed beyond 2006, their value will surely increase.

Recommendation. Sensor technology should be leveraged immediately to develop sensor systems to improve a soldier's ability to discriminate among friends, foes, and noncombatants in all terrain and all weather conditions at much greater battlefield ranges.

Materiel Alternatives Against Mounted Targets

The committee considered nine alternatives to current mixed systems, using the Volcano M87 as a baseline for comparison. The committee was also provided with descriptions of systems under consideration by DOD as part of the Track III search for alternatives. None of these systems had been developed enough to be assessed, although several did appear to be promising. Because of the need to protect proprietary information, none of them is described here.

Under current policy, no fully equivalent alternative to mixed systems is likely to be available by 2006. Other than the Track III search for an alternative, little is being done that could lead to the fielding of a satisfactory alternative. The Hornet/WAM, with its large lethal radius and antihandling device, could replace most of the tactical functions currently provided in mixed systems but has no remote delivery capability. If a satisfactory remote delivery capability could be developed by 2006, the Hornet/WAM appears capable of performing the mixed-minefield mission satisfactorily.

Recommendation. Promising Track III concepts should be developed into weapon system programs. The development of any of these concepts by the 2006 deadline, however, would require that considerable additional resources be allocated for development and procurement.

Protecting Antitank Mines

The DPICM with random-delay fuzing (popcorn) did not score as well as other alternatives for protection against a mounted target. Although random detonations might deter a dismounted enemy, they could also result in friendly casualties and the fratricidal explosion of nearby AT mines, as well as alert the enemy to the location of the minefield. Random detonations would not provide an early warning of ground attack as would a sensor or activated lethal or nonlethal APL alternative. One advantage of DPICM/popcorn is that it could be used on its own as an area-denial weapon.

Nonlethal Alternatives in Mixed Systems

CLADS, a joint Army-Marine Corps nonlethal program currently on hold, is designed to be remotely delivered and could be deployed with AT mines to protect them. CLADS was evaluated against the mixed system baseline of the Volcano M87, both as a weapon launched alongside AT mines and as part of a mixed system. The committee also considered an adaptation of the current Volcano system, which would include one CLADS nonlethal munition in the same canister as five Volcano AT mines, creating a mixed system with a nonlethal APL component. The advantage of this adaptation over stand-alone CLADS is that the APL alternative would be more uniformly dispersed among the AT mines. This mixed system concept, however, has not yet been tested and would require more research and development than the purely nonlethal version. In general, however, CLADS is a promising alternative that may protect AT minefields from dismounted breaches.

Recommendation. The development of nonlethal variants to support antipersonnel landmine alternatives should be emphasized. Funding should be restored and development accelerated for the nonlethal Canister-Launched Area-Denial System (CLADS). The CLADS munition should then be integrated into Volcano (M87A1) canisters to provide a mix of antitank and nonlethal antipersonnel munitions.

Antitank Mixed Systems

SSDMS would consist of hand-emplaced canisters containing a combination of APL and AT mines that would remain in the canister, thus harmless, until a radio command ejected and activated them. Although this system would be an improvement over today's nonself-destructing APL and AT mines, it would also have some disadvantages. The canisters could not be remotely emplaced, meaning that SSDMS could not be used in rapid maneuver situations. In addition, because the system includes APL, it would not be compliant with the Ottawa Convention.

Another mixed system considered by the committee was RADAM, a concept under development by DOD that would combine the existing RAAMS (AT mines) and ADAM (APL), which are now fired separately, into one projectile. This is another contentious issue which is more fully described in Box 6-2.

Recommendation. Until a long-term solution can be developed, the Area-Denial Artillery Munition (ADAM) should be retained in the inventory for use with the Remote Antiarmor Mine System (RAAMS). Production of the Remote Area-Denial Artillery Munition (RADAM) should be halted and funding redirected toward the development of long-term alternatives for mixed systems.

BOX 6-2
Remote Area-Denial Artillery Munition (RADAM)

The use of all pure APL (except the Claymore) has been limited by presidential policy to Korea after 2003. To maintain an artillery-delivered mixed landmine system, DOD has only two options: (1) request a change in presidential policy to allow the continued use of ADAM fired in tandem with RAAMS; or (2) develop RADAM. ADAM is the only artillery deliverable APL in the U.S. inventory. Rather than lose this capability, DOD devised a short-term solution combining ADAM and RAAMS into a single 155-mm projectile to maintain the mixed capability in artillery-delivered scatterable mines until another alternative is developed. The combined RADAM projectile would have the benefits of single-tube delivery and simplified logistics. Otherwise, RADAM would provide no significant military advantage over the combined use of RAAMS and ADAM. In addition, RADAM would not be Ottawa compliant.

The committee concluded that the disadvantages of RADAM outweigh its advantages as an interim solution. The creation of RADAM would take the Ottawa-compliant RAAMS out of the inventory and create a new noncompliant munition. The funds spent creating RADAM could be better spent on the development of an Ottawa-compliant alternative.

Antitank-Pure Mine Systems

In the near term, several existing mixed landmine systems will continue to be capable of destroying tanks and other vehicles without APL components, albeit with an undefined loss of protection from dismounted breaches. Although much testing remains to be done to determine the amount of protection APL provide to AT minefields (see Appendix D), the committee considered the removal of the APL component of two mixed systems, MOPMS and Gator. Their military effectiveness would then be equivalent to that of the current Volcano M87A1 and RAAMS, which are also AT-pure systems. In fact, all four systems were scored similarly, although slightly different military advantages and disadvantages would result from different modes of employment. The AT-pure MOPMS had a higher humanitarian score because a man-in-the-loop would dispense the mines.

One idea developed by the committee was RAAMS enhanced with telemetry (RD-Telemetry), which would involve upgrading the existing projectile that contains AT mines with a subminiature telemetry and communications package that could calculate the precise dispensed locations of mines and communicate the information back to friendly forces. The benefits of RD-Telemetry would include reduced fratricide by improving the situational awareness of friendly soldiers and by more accurate location of the remotely delivered minefields. In addition, if combat commanders were immediately informed of the locations of deployed mines, they could use them just in time and also reduce the enemy's ability to find and clear the minefield. Although much research and development would be necessary to provide more complex communications, the technology might be applied to other submunitions.

The system the committee considered to be the best AT-pure alternative against a mounted enemy available by 2006 is the Hornet/WAM PIP. This two-phased, evolutionary improvement to the existing Hornet/WAM, hand-emplaced AT mine is a comprehensive, fully funded research, development, testing, and evaluation program. The first phase, scheduled to be type classified (accepted for service) by fiscal year 2002, would add a man-in-the-loop to monitor the sensors and partially control the minefield, while the mines continue to operate autonomously. The second phase, scheduled for type classification by fiscal year 2003, would add a more sophisticated sensor, an improved target detection device, and a dual-purpose warhead. In the committee's opinion, the Hornet/WAM PIP's much greater kill radius would provide military advantages over the baseline Volcano M87, forcing an enemy to clear a passage lane more than 200 meters wide for tanks, thus, significantly delaying a dismounted breach.

The disadvantage of the Hornet/WAM PIP is that it is not remotely deliverable, and the PIP does not include this development. Other allied nations have fielded hardened, Hornet/WAM-sized AT mines that can be delivered by MLRS. Hardening the mine and developing a remote delivery capability appears to be an Ottawa compliant, low-risk alternative to current mixed minefields.

TABLE 6-2 Score Sheet for Alternatives Available by 2006

		System Name	HESF	NSD-A w/o switch	Sphinx	Multi-Shot Claymore	BNLM	Taser Nonlethal	WAM PIP	RADAM	RD Telemetry	CLADS	Volcano-CLADS	AT Pure-MOPMS	AT Pure-Gator	Popcorn	SSDMS
		Available for Implementation	06	06	06	06	06	06	06	06	06	06	06	06	06	06	06
		Baseline D = M14/16 M = M87	D	D	D	D	D	D	M	M	M	M	M	M	M	M	M
	A0	**Military Effects Sum:**	**12**	**10**	**4**	**4**	**–3**	**–4**	**7**	**–1**	**–1**	**–1**	**–2**	**–2**	**–2**	**–3**	**–3**
Mounted	A1	Enhances effects of close and deep friendly fires							2	1	1	–1	–1	0	1	–1	0
Mounted	A2	Has multiple methods of delivery							–1	–1	0	0	0	–1	–1	0	–1
Mounted	A3	Provides a range of effects that inhibit mounted and dismounted maneuvers							1	0	–1	–1	–1	–1	–1	0	0
Mounted	A4	Resists full spectrum of enemy breach methods, including dismounted methods							1	0	–1	–1	–1	–2	–2	2	0
Mounted	A5	Provides early warning of ground attack							1	–1	–1	0	0	0	0	–1	0
Mounted	A6	Is safe for friendly forces							0	0	0	0	0	0	0	0	0
Mounted	A7	Is effective in all types of terrain and weather							0	0	0	0	0	0	0	0	0
Mounted	A8	Poses minimal residual hazard to own forces and noncombatants after military conflicts							1	0	0	1	0	0	0	0	0
Mounted	A9	Is difficult to detect by enemy forces							–1	0	0	0	0	0	0	–2	0
Mounted	A10	Poses minimal risk of fratricide							2	0	2	1	1	1	1	0	0
Mounted	A11	Has modifiable effects for mounted and/or dismounted threat							1	0	0	0	0	0	0	0	0
Mounted	A12	Has controllable activation/deactivation and duration before and after installation							2	0	0	0	0	2	0	0	0
Mounted	A13	Is effective in nuclear, chemical, and biological environments							–1	0	–1	0	0	–1	0	0	–1
Mounted	A14	Is easy and efficient to distribute							–1	0	0	0	0	0	0	–1	–1
Dismounted	A15	Can delay, disrupt, and/or canalize enemy movement/maneuvers	2	2	1	1	–1	–1									
Dismounted	A16	Denies enemy access to terrain/facilities (including short- and long-term deterrent for boundaries and DMZ areas)	2	2	1	1	–1	–1									
Dismounted	A17	Enhances effects of friendly-force weapons, obstacles, and munitions (including AT mines)	2	2	1	1	–1	–1									
Dismounted	A18	Generates exploitable delays and opportunities (fixes or contains enemy)	2	2	1	1	–1	–1									
Dismounted	A19	Generates detection, alert, and/or early warnings	2	1	–1	–1	0	–1									
Dismounted	A20	Facilitates classification of target	2	1	1	1	0	0									
Dismounted	A21	Produces desired effects on enemy forces (nonlethal to lethal)	0	0	0	0	0	0									
Dismounted	A22	Reduces casualties/risk for U.S. and/or allied forces	1	1	1	1	2	2									
Dismounted	A23	Deters pursuit to facilitate breaking of contact under pressure	–1	–1	–1	–1	–1	–1									
	B0	**Humanitarian Concerns Sum:**	**6**	**5**	**5**	**5**	**2**	**2**	**5**	**1**	**3**	**4**	**3**	**4**	**2**	**2**	**2**
	B1	Reduces postconflict residual hazard (CCW)	1	1	1	1	1	1	1	1	1	1	1	1	1	1	1
	B2	Location can be recorded (CCW)	1	1	1	1	0	0	1	0	1	0	0	0	0	0	1
	B3	Detectable to facilitate removal (CCW)	1	0	0	0	0	0	0	0	0	0	0	0	0	0	0
	B4	Discriminates between combatants and civilians (Ottawa)	2	2	2	2	0	0	2	0	0	2	2	2	0	0	0
	B5	Does not explode on presence, proximity, or contact of a person (Ottawa)	1	1	1	1	1	1	1	0	1	1	0	1	1	1	0
	C0	**Overall Technical Risk (C4ISR, etc.)**	**–2**	**–1**	**–1**	**–2**	**–1**	**–1**	**–2**	**–1**	**–2**	**–1**	**–1**	**–2**	**–2**	**–2**	**–2**
	D0	**Requires Change in Tactics/Doctrine**	**Y**	**Y**	**Y**	**Y**	**Y**	**Y**	**Y**	**N**	**Y**	**Y**	**Y**	**Y**	**Y**	**Y**	**Y**
	E0	**Cost Sum:**	**–4**	**–2**	**–2**	**–4**	**–2**	**–2**	**–4**	**–1**	**–3**	**–2**	**–5**	**–4**	**–4**	**–4**	**–4**
	E1	Research and development	–2	–1	–1	–2	–1	–1	–2	–1	–1	–1	–3	–2	–2	–2	–2
	E2	Procurement	–3	–2	–2	–3	–2	–2	–3	–1	–3	–2	–3	–3	–3	–3	–3

NOTE: Acronyms are defined on p. xv in the front matter.
Key: D = dismounted, M = mounted, n/a = nonapplicable.

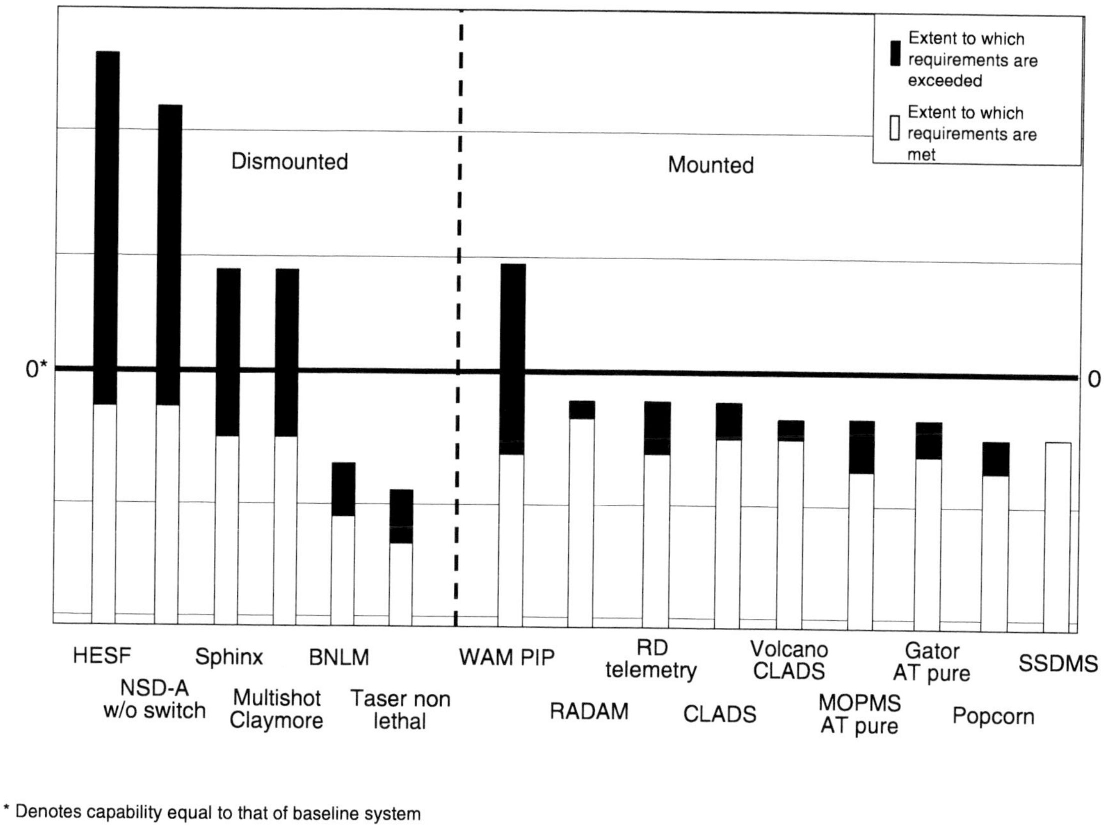

FIGURE 6-1 Military effectiveness of alternatives available by 2006 based on qualitative scoring by the committee.

Recommendation. The feasibility, cost, and schedule of providing a remote delivery option for Hornet/WAM should be investigated. Shock hardening of the mine to withstand the impact of remote delivery appears to be an Ottawa-compliant, low-risk solution to current mixed minefields.

Summary

The criteria and scores are shown in Table 6-2. The details of how these scores were derived can be found in Chapter 4.

Figure 6-1 is a graphical summary of the scoring. In keeping with the Statement of Task, this graph shows only the relative military effectiveness of candidate systems without regard to cost, risk, or humanitarian factors. Each bar on the graph is a composite. The lower portion (white) shows the degree to which each system meets the military effectiveness requirements in comparison to the baseline system. If the candidate system meets all of the requirements at least as well as the baseline system, the score is 0. If it is less effective in any requirement, the score is less than 0. The upper portion (dark shading) of the bar shows capabilities that exceed those of the baseline system.

These graphs use the methodology described in Chapter 5. In general, if the total height of the bar is high, the system is likely to be militarily effective. If the value of the lower portion of the bar is near 0, the system meets most of the military requirements. If the lower bar is much lower than 0, the system probably has significant differences from the baseline mine and will not perform some desired functions. However, that system may still be militarily effective if it performs some functions much better than the baseline system. Because the scoring criteria were not weighted, these graphs should be used only for assessing trends and making qualitative comparisons.

7

Alternatives Potentially Available After 2006

> We can transform today in a time of peace and prosperity. Or we can try to change tomorrow, on the eve of the next war, when the window has closed, our perspective has narrowed, and our potential is limited by the press of time and the constraints of resources. (Shinseki, 2000)

OVERVIEW

The near-term and midterm alternatives described in Chapters 5 and 6 are purposely conservative, which has little to do with available technology and much to do with the time required for DOD's decision, program development, procurement, and acceptance procedures for new weapons. In other words, without fast-track development and procurement procedures, even suggested systems using available technology and well-known assembly practices may not be available for service use by 2006. In this chapter, which describes alternatives that might be available after 2006, the committee is under no such constraints. Therefore, available and emerging technologies are more liberally considered for use in potential APL alternatives.

Advanced Technologies

Advanced technologies will have a profound effect on the capabilities of U.S. forces. The rapid emergence of new technologies will create opportunities after 2006 for the development of future systems that can outperform today's APL and be compliant with Ottawa. Alternative systems that separate the sensor from the shooter will provide powerful new capabilities, particularly in sensing and command and control, which could not only replace APL, but could also reinforce the information superiority concept in *Joint Vision 2010* and *Joint Vision 2020.* Information systems to manage battlefield operating systems, especially interactive communication with deployed mines, could be one of the most significant features of future mine warfare. Future systems could provide precise locations and operational status reports for all types of mines in near real time. Cost may be the controlling factor, but the capability seems to satisfy a key requirement of the CCW.

Elements of technology development expected to be pertinent to APL alternatives include advanced intelligence, sensors, and reconnaissance capabilities; new weapon systems and munitions; and integration through networking. The Army Research Laboratory's *Annual Review* publishes many descriptions of projects in support of a future digitized battlefield, including ideas for advanced sensors, signal and image processing, displays, information distribution, visualization, modeling, simulation, vehicles, armor, and munitions (U.S. Army, 1998a). Publications by laboratories in other military services include similar scenarios.

In addition to retaining the militarily desirable characteristics of current APL, future systems may satisfy new requirements, including the ability to distinguish among friends, foe, and neutral parties rapidly and reliably; easy recovery after hostilities; and having benign effects on the environment.

Civilian applications are expected to continue to lead the way in communications and information technology. Indeed, DOD is no longer the driving force behind research and development and applications of many technologies in the United States. Therefore, the use of commercial off-the-shelf hardware and software for military communications and information systems seems certain to increase.

Surveillance and Targeting Sensors

The military requirement for comprehensive surveillance of minefields and the necessity of accurately targeting lethal or nonlethal components comes at a propitious time in sensor development. With emerging technologies, nearly everything about a battlespace can be known, and anything in it can be hit. The development of sensor technology is characterized by the following trends (NRC, 1997):

- continuing decreases in size and cost, as MEMS evolve into nanoelectronic systems limited only by the physics of the interfaces
- migration of the analog-to-digital conversion to the front end of the sensor, leaving only those analog elements absolutely necessary for interfacing with the physical phenomenon to be sensed (e.g., microwave, low-noise amplification; filters and power amplifiers; fiber-optic transducers; MEMS transducers, etc.)

- increasing applications of computer processing as gigaflops grow to teraflops and then petaflops[1]
- development of monolithic smart sensors that combine sensing transducers, analog-to-digital conversion, digital-signal processing, communications input and output, and, perhaps, power conditioning on a single chip (could lead to the development of very small, very smart sensor systems and weapons, including affordable smart bullets)
- integration of autonomous, mobile, communicating sensors that can cooperate to function as single, high-level metasensors

Recommendation. The development of sensor-net technology should be pursued aggressively and eventually incorporated into a fully militarized, deployed system characterized by networking, strong detection and tracking capabilities, robustness, low power consumption, low cost, covertness, low probability of intercept, easy deployment, and disposability.

Recommendation. Investments already being made in new technologies for other purposes should be leveraged and applied to the search for alternatives to antipersonnel landmines.

Network-Centric Battlespace

With major advances in communications, all elements in and supportive of a battlespace are being linked in near real time, making possible what has come to be called network-centric warfare. The new systems will involve sensors, communication and communication relays from satellites, manned and unmanned aerial vehicles, sea-based and ground-based mobile platforms, and ground stations.

In most cases, the two-way communication link between the surveillance element of a future minefield alternative and a remote operator would not require line of sight, would be secure, and would be capable of working with network-centric architecture. Its primary purpose would be to enable a remote operator to evaluate mounted and dismounted intrusions into the minefield, distinguish enemy from friend and noncombatant, and control the actions of the lethal and/or nonlethal elements. Its secondary purpose would be to contribute to overall situational awareness in the battlespace.

During the course of this study, a good deal of discussion was focused on the ability (or inability) of a remote operator to remain alert and to control the minefield during an intrusion, possibly under extreme combat conditions. An advantage of integration of the operator-to-minefield communication link with the network-centric architecture is that the operator would not have to be in the combat zone.

[1] Flops is a unit of computer speed equal to one floating-point arithmetic operation per second. Giga is 10^9 (one billion), tera is 10^{12} (one trillion), and peta is 10^{15} (one thousand trillion).

Energetic Explosives

Until 1940, military explosives such as TNT and mercury fulminate generated an energy release of approximately 1 kilocalorie per gram. During World War II, advanced compounds, such as nitramines, were increasingly introduced. During the Cold War, the energy content of explosive compounds continued to increase slowly and steadily. Today the energy content of military explosives is approaching 4 kilocalorie per gram.

Research supported by all of the armed services in the past 15 years indicates that the energy release of military explosives, acceptably desensitized, is likely to be doubled in the foreseeable future. This improvement will allow a significant reduction in the size and weight of explosive devices, as well as further advances in special effects. An increase by a factor of two could also have a profound effect on logistics and combat effectiveness.

Minefield Deliverability

Current pure and mixed minefields can be emplaced by hand, ground vehicle-mounted dispensers, helicopters, fixed-wing aircraft, artillery, and missiles. As lighter forces are used in more situations, tactical minefields that can be quickly and remotely emplaced and quickly removed through command detonation and/or retrieval of the more expensive and reusable components will be necessary. The trend in remote mine delivery modes will be toward artillery, missiles, fixed-wing, and rotary-wing aircraft.

The potential size reduction of explosive devices combined with alternative systems to replace minefields with small alerting sensors will make the timely, remote emplacement of mines easier. Extended range guided munition-like artillery rounds and Army TACMS missiles will provide effective delivery modes. However, the accurate emplacement of the minefield surveillance component, particularly if it is combined with an alternative system, could complicate delivery. For example, the surveillance/kill system could be delivered separately by helicopter or Osprey (V-22) and hand emplaced by the crew once the location of the minefield had been determined, its boundaries known, and the best site for the surveillance/kill systems determined. At some point, precision emplacement of components using GPS, combined with digital maps, will make remote emplacement of the surveillance/kill system feasible.

Location and Precision Emplacement

For precision weapon delivery, modeling, simulation, and operational planning, the U.S. military has long had a requirement that maps be accurate to within 30 meters. In the current geopolitical environment, such maps are required for most of the earth's land surface. The recent Shuttle Radar Topography Mission now promises to provide digital maps of more than 80 percent of the earth's land surface and 95 percent of populated areas with data taken at 30-meter intervals and an accuracy of ± 15 meters in the vertical and ± 30 meters in the horizontal. These maps, combined with GPS guidance and position reporting, will profoundly influence all aspects of military activity, including site location, precision emplacement, and precision boundary marking of future minefields.

Vulnerabilities

New technologies provide not only new improved capabilities, but also new or additional vulnerabilities. This will certainly be true of operational interfaces between communication and sensor systems, as well as the more general battlefield C4ISR systems. The very large volume of information now available through the C4ISR systems must be balanced against the tactical and operational requirements of the warfighter who must respond to changing situations with great immediacy and reliability, avoiding, as much as possible, information traffic jams, delays, and disruptions that often affect communications systems. Systems that take advantage of future C4ISR network capabilities will probably rely on radio-frequency communication links, which can be vulnerable to jamming and other countermeasures. Because the cost of ensuring against jamming in systems used for every munition may be prohibitively high, some loss of system performance as a result of active countermeasures should be expected.

Defense Advanced Research Projects Agency

DARPA manages and directs selected basic and applied research and development projects. The agency pursues research and development for which both risks and payoffs are very high, and successes may lead to dramatic advances in traditional military roles and missions, as well as dual-use applications. DARPA has been involved in the search for alternatives to APL, directly through research and development on Track II and indirectly by research and development on a variety of devices that can increase the likelihood that future alternatives, particularly sensor-related ones, will be developed successfully.

Representatives of DARPA met with the committee on several occasions, and small groups of committee members visited the main DARPA facility twice. The following descriptions include a variety of ongoing DARPA programs that may affect the development of future APL alternatives.

Affordable, Moving-Surface-Target Engagement Program

This program conceptually leverages recent advances in sensor technology for the development of an affordable, precise means of identifying and destroying a moving surface target. The fundamental concept to be investigated is a network of two radar systems, ground moving targets indicating (GMTI) radar and synthetic-aperture radar (SAR), to provide precision fire-control tracking of moving surface targets. The network would update precision-guided munitions in flight for precise engagement of moving surface targets (Grayson, 2000).

The goal of the program is to develop, investigate, and evaluate technologies that could lead to affordable architectures for destroying specific moving targets on land, the littorals, and water. The focus is on weapon system technologies that would enable precision, affordable, all-weather engagement of a wide range of moving surface targets, both on land and at sea. Research and development will focus on the use of netted GMTI/SAR sensors to provide precision fire control of inexpensive, nonsmart weapons (DARPA, 2000c).

SensIT Program

The SensIT Program is founded on the concept of a networked system of inexpensive, pervasive platforms that combine multiple sensor types, reprogrammable general-purpose processors, and wireless communication (Kumar, 2000). The multiple-sensor module might combine optical, acoustic, triple-axis, seismic, magnetic, moisture, pollution, poison, organic pressure, temperature, acceleration, and physiological variables. The goal of the SensIT Program is to create an interface between the physical world and cyberspace.

Current information systems use human input or computer-generated data. Future systems will build on continuous streams of real-world physical data to create a "virtual" supercomputer, miniaturized and distributed into the environment, with each node computing and collaborating to "see" into its sensor region. The mission of the SensIT program is to develop all necessary software for networked microsensors (DARPA, 2000d).

Microunattended Ground Sensors Program

The goal of this program is to develop a distributed network of miniature unattended ground sensors (UGSs) based on acoustic, magnetic, seismic, meteorological, and imaging technologies, and advanced fusion algorithms for tactical use (Carapezza, 2000). These sensors should be low-power miniature imaging and nonimaging variants. The long-term objective is to support the pursuit of time-critical mobile targets, combat vehicles, and dismounted soldiers. Miniature UGSs could be used singly or in networks to provide a local, in-situ detection, tracking, and identification capability at high-value manmade facilities or at choke points in denied areas (DARPA, 2000e).

Future Combat System Program

The future combat system (FCS) will be a multifunctional, multimission, reconfigureable system of systems designed to maximize joint interoperability, strategic transportability, and commonality of mission roles, including direct and indirect fire, air defense, reconnaissance, troop transport, countermobility, nonlethal options, and command and control on the move. The goal of the program is to develop a network-centric advanced force structure that will be overwhelmingly lethal, strategically deployable, self-sustaining, and highly survivable in combat through integrated command and control capabilities that provide unsurpassed situational awareness for commanders at all levels (DARPA, 2000a).

Nonlethal Alternatives

Nonlethal devices of the future might use real or virtual images. Making large images appear by projection or reflection of small objects is not difficult if the viewing point and the ambient light level are controllable. However, for the illusion to extend over a large space, to be visible from various angles, and to be viewable in daylight, the problem is much more complicated. Images are, after all, nothing more than a directed configuration of photons. The more extensive the image is spatially, the more illuminating power it requires; hence, the availability of a power source on the battlefield may be a problem. Climate, vegetation, and terrain may also disrupt or degrade images.

A static image is not likely to arouse fear or dread for an extended period of time. With the advent of inexpensive, robust scanning technology, along with lightweight mirror materiels, a plasma point could be created with one or more focused beams swept over an area to create a more realistic image. Although a single laser creates only a single plasma point, hysteresis in the human eye makes the image appear constant and solid. (For instance, on a TV screen, a single electron beam scans the phosphor dots on the inside of the screen many times per second, exciting the appropriate dots one at a time to create an image that the eye perceives as constant and whole.)

The use of movement is yet another futuristic possibility. "Seasickness" is familiar to most people, if not directly, then by empathy or observation. Small amplitude vertical vibrations of 0.5 Hertz are known to create this effect. If enemy troops and equipment must cross bridges or other solid manmade structures, it may be possible to deploy a device that generates small amplitude movements at the nauseogenic frequency, thus incapacitating troops with seasickness. This device could provide in such situations a more effective delay than APL (Haseltine, 2000).

As discussed earlier, nonlethal variants have certain drawbacks: (1) nonlethal systems could cause inadvertent fatalities; (2) they are likely to be less of a deterrent to a determined enemy and may even be interpreted as weakness; and (3) even if nonlethals confuse an enemy initially, he is unlikely to make the same mistake twice. Nevertheless, in light of the increasing frequency of peacekeeping operations, the development of nonlethal variants to support APL alternatives should be a high priority.

Recommendation. Several other technologies or systems already under development for other purposes should be considered as potential components of long-term alternatives to antipersonnel landmines, including unmanned air and ground vehicles, directed-energy weapons, battlefield sensory-illusion devices, passive transponders (e.g., tags), and other lethal and nonlethal systems.

MATERIEL ALTERNATIVES

After 2006, improvements in the tactical effectiveness of existing or proposed remotely delivered AT landmines ought to be technologically feasible, which could eliminate the need for mixed systems. Future systems that separate the sensor from the shooter could be improved by multiple means of remote deployment and resistance to countermeasures through signature reduction and other techniques. Track III programs, like the Track I initiative, will require concentrated effort and stable funding. In the long term, the emergence of new technologies, such as the ability to distinguish accurately between combatants and noncombatants, could lead to the development of systems that can outperform today's APL.

The most promising alternatives received high scores on both military effectiveness and humanitarian criteria, which reflects the greater battlefield awareness provided by advanced technologies. The same alternatives received low scores for technical risk and economic criteria because they tend to be conceptual and on the cutting edge of technology.

Table 7-1 shows systems that might be available sometime after 2006. The table also describes their principle characteristics. Full descriptions and brief written assessments follow, as well as a table measuring each alternative against the criteria described in Chapter 4.

For Use Against Dismounted Threats

Radio/Radar Sensor Munition System

Source: Committee on Alternative Technologies to Replace Antipersonnel Landmines

The Radio/Radar Sensor Munition System (RRASMS) would consist of four parts: a sensor and communications unit that would function as a hub for a section of the denied zone; an overwatch controller to control the modular munitions; the modular munition; and an electronically programmable radio.

The radio would perform three functions:

- self-location and munitions location using either GPS or multilateration with other units
- communication to the overwatch controller unit with a soldier/operator in conjunction with other sensor and communications units
- multistatic radar to detect and track human intruders

TABLE 7-1 Alternatives Potentially Available After 2006

System Name	APL/ AT/ Mixed Non-Mine	Self-destructing/ Self-deactivating	Lethal/ Nonlethal	Ottawa Compliant[a]	Dismounted Enemy		Mounted Enemy	
					Remotely Delivered	Hand Emplaced	Remotely Delivered	Hand Emplaced
Radio / Radar Sensor Munition System (RRASMS)	APL	Y	L	Y		X		
Unmanned Remote Ambush System (URAS)	APL	Y	L	Y		X		
Tags / Minimally Guided Munitions **Track II**	n/m	n/a	n/a	Y	X			
Laser Radar Directed Machine Gun (LDMG)	n/m	n/a	L	Y		X		
Distributed-Sensor Antipersonnel "Minefield"	n/m	n/a	L	Y	X			
Distributed Web-Sensor Complex (DWSC)	n/m	n/a	n/a	Y			X	
Raptor	AT	Y	L	Y			X	
RAAMS Enhanced with Telemetry and Sensor Package (RD Sensor)	AT	Y	L	Y			X	
Remotely Delivered Hornet/WAM (RD-WAM)	AT	Y	L	Y			X	
Self-Healing Minefield **Track II**	AT	Y	L	Y			X	
BAT Antiarmor Mine (BATAAM)	AT	n/a	L	Y				X
Early Warning Subsystem for Remotely Delivered AT Minefields (EWSS)	n/m	n/a	n/a	Y			X	
RAAMS with Nonlethal Capability (RAAMS-NL)	Mix	Y	N/L	Y	X		X	

[a]The committee used the definition found in the Ottawa Convention to determine whether a system would be Ottawa compliant.

The modular munition units would be connected remotely to the sensor and communications unit via trip wire/communication lines. Remotely deployable versions might use radio-frequency links. Modular munition units would come in three basic types:

- warning devices, such as flashers or sirens
- nonlethal deterrents, such as flash/bang units and malodorants
- lethal devices, such as small fragmentation grenades

The overwatch control unit, a computer terminal with a radio, would perform three functions:

- display a situational awareness map to the soldier/operator showing the geometry of the munitions in the denied zone
- sound an alert and display the track of an intruder
- allow the soldier/operator to command a lethal response when necessary

To create a denied zone, RRASMS would be deployed in the following way. The soldier/operator would place sensor and communications units in the denied zone approximately 50 meters apart. He would then attach as many as 32 modular munition units to each sensor and communications unit in dispersed locations. In general, warning devices would be placed farthest forward in the denied zone; nonlethal devices would come next; and lethal devices would be placed behind the others.

Once the field was activated, the sensor and communications units would go through an initialization process to determine their locations and the locations of the modular munition units. Each sensor and communications unit would periodically transmit a radar pulse, and other sensor and communications units would listen. If an intruder were detected, the overwatch control unit would be alerted. Depending on the situation, the sensor and communications unit could autonomously activate warning and/or nonlethal devices. After assessing the track of the intruder(s) and the response to the nonlethal devices, the soldier/operator could command a lethal response.

Advantages

- This system would be compliant with the CCW Amended Protocol II and the Ottawa Convention.
- With an electronically programmable radio, the RRASMS could communicate with a wide variety of radios on the battlefield, could receive GPS signals, and provide guidance updates to incoming ordnance.
- With a radar mode, it could provide all weather, day/night sensing and tracking of human intruders and would have some foliage penetration capability.

- As a backup, the wires connecting the modular munition units to the sensor and communications unit could also serve as trip wires for sensors.
- The soldier/operator would have the flexibility to attach any mix of warning, nonlethal, and lethal modular munition units to the system.
- The system would give the operator the time and information necessary to determine if a lethal response is necessary.
- The system could be enhanced by providing devices to friendly soldiers that would identify them as friendly and would disable lethal responses in their vicinity, thus allowing them free passage through the denied zone and avoiding fratricide.
- RRASMS could be used in conjunction with covert tags on enemy soldiers for longer range tracking.

Disadvantages

- RRASMS would have significant technical risk and development costs and would require successful development of electronically programmable radios, which would significantly reduce the risk and cost of adding the radar function.

Unmanned Remote Ambush System

Source: Committee on Alternative Technologies to Replace Antipersonnel Landmines

The ambush has been an effective tactic throughout the history of warfare and will continue to be effective in certain types of future conflicts. With modern technology, ambushes could be operated without on-site personnel.

The mine used in the unmanned remote ambush system (URAS) concept could be any of a number of current APL mines. Claymores, modified for timed self-deactivation and command detonation, would be most appropriate for ambushes and would be CCW and Ottawa compliant.

URAS would require small cameras for discriminating between friend and foe. The US Army Night Vision and Electronic Sensors Directorate is working on the development of an uncooled infrared camera (a the forward-looking infrared radar, microcamera [UL3]) that is about 5 centimeters long and 6.4 centimeters square, weighs 70 grams, and requires 540 microwatts at 3.5 volts. The UL3 can detect a walking man at ranges of 250 to 700 meters, day or night, depending on the angle of view. To conserve power, an acoustic instant wake-up for the camera is available. Also available, if required, is an eye-safe laser illuminator for better target identification. DARPA is developing a television camera of similar size, capable of projecting its image onto the upper quadrant of a specially equipped pair of glasses or goggles (if the operator is moving).

The dispersal of Claymores at the ambush site would depend on the nature of the terrain and the anticipated size and dispersion of the intruding force. The camera and laser illuminator could be colored to blend in with the environment and attached to a tree, rock, or stake to provide the proper field of view. The preferred communication link would be an aircraft or satellite so that distance and terrain would not matter. A small broadcast terminal and disk antenna would be required for this communication mode.

In operation, the first transmission from the ambush would be accompanied by a sound and/or flashing light to alert the remote operator. Upon identification of the target, the operator would detonate the Claymores at the proper distance between mines and intruders. The Claymores might be given different firing codes for staggered firing. Antihandling features might be added, as needed.

URAS is covered by doctrine governing ambushes and APL minefields. All of the major components of URAS are available or in an advanced stage of development. The components would have to be merged into a system and a two-way communication package assembled.

Advantages

- URAS would be CCW and Ottawa compliant.
- The concept would not require on-site personnel.
- No friendly lives would be placed at risk during the ambush.
- The URAS poses little logistics burden and could be easily and quickly emplaced.
- URAS provides firing versatility for maximum effectiveness.

Disadvantages

- URAS would require a reliable two-way communication link with one-way imagery.
- Aircraft might be necessary for the communication link. However, the aircraft would probably be used for multiple purposes.
- Camera performance may be degraded by adverse atmospheric conditions (e.g., heavy rain, fog, etc.).

Tags/Minimally Guided Munitions

Source: DARPA Track II (Altshuler, 1999)

The Tags/Minimally Guided Munitions concept is under development at DARPA. The agency has already initiated research and development on required technologies for this system. Preliminary demonstrations of tag attachment have met with some success. However, it is extremely unlikely that the system could be available by 2006.

In this system, small burr-like transmitters would be affixed to the clothing of enemy soldiers as they traverse a field (called a "tag-field"). The method of attaching the tags is still under development; a "lawn-dart" and a "brier" are being considered. The tags, anticipated to be smaller than

0.5 centimeters in all dimensions, would alert a man-in-the-loop with motion sensors or equivalents. Tags would provide one- or two-way communication over a short transmission range of less than 100 meters (longer transmission would require a relay network). The man-in-the-loop would launch munitions, which would be target-oriented rather than area-oriented and would home in on an individual tag or cluster of tags, making small in-flight course corrections as necessary. The course correction capability would have to be greater than the distance the target could move in the interval. Tag lifetime would range from minutes to hours.

Effective tags will require millimeter-sized transmitters and antennae. The power source for the lifetime of tags will also require further work (thick or thin battery technology), as will delivery of the tags and their adhesion to the target. Other issues that would have to be addressed include: packaging tags; delivering tags; recognizing and discriminating targets; reducing vulnerabilities to countermeasures; extending the transmission range; and developing repeaters (a multitiered communication system) to ensure that communications reach the command center.

Research and development on the munitions will have to address the following issues: homing technology using a radio-frequency signal; lowering the cost and increasing the sensitivity of the receiver; flight control/flight errors (precision strike) and time of flight; and the overall efficiency of operation.

Integrating tag and munition technologies to ensure reliable operation and include a man-in-the-loop will involve tactical changes based on studies of the behavior of individual soldiers and units. An overall cost analysis and technology implementation routine would also have to be developed.

Advantages

- This system is envisioned to be compliant with the CCW Amended Protocol II and the Ottawa Convention.
- The system would improve situational awareness.
- Not many tags would be required for the system to be effective (modeling shows that only one-third of the enemy population would have to be tagged).
- Tags would be particularly effective for protecting flank positions and preventing infiltration by small groups.
- Environmental and post conflict effects would be minimal.

Disadvantages

- It is not clear how the munition would differentiate among moving tags and home in on a specific target.
- Communications might be jammed as a consequence of in-flight confusion.

Laser Radar-Directed Machine Gun

Source: Committee on Alternative Technologies to Replace Antipersonnel Landmines

The Laser Radar (LADAR)-directed machine gun (LDMG) would use laser radar (or other means) to maintain surveillance over a denied zone and for precision aiming of an automatically aimed machine gun. The machine gun would fire two types of munitions: nonlethal rubber bullets and lethal explosive/fragmentation rounds. The gun would have an antihandling mechanism that would disable it in the event of enemy capture. An overwatch control unit would have to be activated before a lethal response was initiated.

The LDMG would consist of four units. The first unit, the LADAR surveillance sensor and fire control unit, would use laser radar to create a three-dimensional picture of the denied zone out to a range of about 500 meters and across an angle of about 60 degrees. If a change in the background were detected, the unit would zoom in on that area to identify the intrusion. The unit would also act as a very precise fire-control system for both nonlethal and lethal responses from the machine gun. The second unit, a machine gun (the objective individual combat weapon), would be a low-recoil system that could shoot both 5.56-mm bullets from its top barrel and exploding 20-mm projectiles from its bottom barrel. For the LDMG, the top barrel would shoot nonlethal rubber bullets. An add-on, increased ammunition feed capability would be developed. The third unit, the gun cradle, would be a tripod with servomotors that could aim the gun based on inputs from the fire-control system. The fourth unit, the overwatch control unit, a computer display that would receive alerts and images via radio from the LADAR, would allow the soldier/operator to determine if the intrusion required a lethal response.

The LDMG would first be used to create a denied zone in the following way. First, one or more LDMGs would be set up behind the denied zone. In general, two or more LDMGs would be used to obtain crossing fires. The LADAR would scan the denied zone to establish the background image. The soldier/operator would then proceed to his post and test radio connectivity with the LDMG. If an intruder were detected, the LDMG would send an alert and image to the soldier/operator and might respond autonomously with nonlethal rubber bullets from the 5.56-mm barrel. The soldier/operator would determine if a lethal response were required. A lethal response fired from the 20-mm barrel would be a projectile that would explode just above the location of the intruder based on range information from the LADAR. To prevent capture, the surveillance/kill system would be equipped with an antihandling device and rigged for timed self-destruction.

Advantages

- This system would be compliant with the CCW Amended Protocol II and the Ottawa Convention.

- The LDMG would be a tireless area-denial sentry with the flexibility to use either nonlethal or lethal responses.
- This system might have multiple uses.

Disadvantages

- The LDMG would be very bulky and would have high electrical power requirements.
- The LDMG would be less effective in rugged or foliated terrain, in adverse weather conditions, or in the presence of smoke.
- The active sensor might reveal its position.
- The surveillance/kill system would have to be hand emplaced.

Distributed-Sensor Antipersonnel "Minefield"

Source: Committee on Alternative Technologies to Replace Antipersonnel Landmines

This system, which could be used either against dismounted targets or for protection of an AT minefield, would have separate sensor and kill components that would not be co-located. The sensor component, about the size of a tube of Chapstick, would consist of a dismounted sensor (pressure, seismic, or tremble switch, with pressure preferred) and a short-range radio-frequency communicator. These small, rugged, inexpensive sensor packages could be distributed by air, missile, artillery, or hand. Upon activation by an intruder, the sensor would emit a single radio-frequency pulse that would alert the kill component. All sensor packages in a given field would use a unique coded pulse to reduce the chance of spoofing. However, to allow for reuse, the kill component would be set to respond to any of the allocated codes.

The kill component would be a .30-caliber or .50-caliber machine gun mounted on a tripod. To ensure stability, concealment, and a clear field of view, the kill component would be emplaced by hand on the periphery of the sensor field. The system would consist of both optical and infrared sensors for day/night surveillance and would be able to transmit its field of view to a remote operator.

Upon receiving a signal from any sensor in the field, the kill component would be switched on, rise to its surveillance position, transmit its picture to the remote operator for identification of friend or foe, and await permission to fire. Once permission was received, the kill system would become autonomous. The gun would fire only when the target was centered in the cross hairs of the aiming system. The system would be able to train across the entire sensor field and engage multiple targets in rapid succession. The operator could interrupt firing at any time.

The sensor package would use state-of-the-art technology and would not require research. The surveillance system would also use available infrared and optical technologies. The capability of firing only when the target was centered might involve pattern recognition technology or the use of tags now being developed by DARPA for the robot sniper and Land Warrior programs. To prevent capture, the surveillance/kill system would be equipped with an antihandling device and rigged for timed self-destruction.

Advantages

- This system would be compliant with the CCW Amended Protocol II and the Ottawa Convention.
- The most expensive part of the system would be recoverable.
- With recovery of the kill component, no explosives would be left on the battlefield.
- The system would have a significantly higher probability of being effective against multiple intrusions than a typical AP minefield.
- The surveillance/kill system would present a very small target.

Disadvantages

- Removing either the remote operator or the kill component could defeat the system.
- The surveillance/kill system would require research and development and would be expensive.
- The surveillance/kill system would have to be hand emplaced.

For Use Against Mounted Threats

Distributed Web Sensor Complex

Source: Committee on Alternative Technologies to Replace Antipersonnel Landmines

The concept of the Distributed Web Sensor Complex (DWSC) is based on an Army science and technology program intended to enable a commander to take advantage of high-tech sensor and communications technology by providing near real-time situational awareness of the extended battlefield. This approach is intelligence oriented, rather than minefield oriented. The essential aspect of the DWSC is the rapid, remote placement of numerous, inexpensive, expendable sensors by artillery or air. Depending on the situation, emplacement could range from relatively close to friendly positions (within a few kilometers) to extended ranges. Any combination of acoustic, magnetic, seismic, motion, infrared imaging, video, biological or other sensors capable of identifying signals from vehicles and/or humans could be used.

Once sensors had landed on the ground, they would be activated by an accompanying gateway, which would also establish communications with and identify the location of each sensor. The gateway would also establish communication with nearby gateways, thereby creating a large web of sensors that could cover the entire front of a deployed unit (several kilometers). Low-power sensors, such as acoustic sensors, would begin functioning. Upon recognizing certain sensory inputs, these low-powered sensors would signal the gateway to "wake up" the more power-intensive sensors, such as infrared imaging, that had been in a stand-by, power-saving mode. The gateway could be programmed to sort through sensor inputs and look for combinations (target templates) to define an intruder. This information would be sent by communication links back to an operations center where sensor intelligence could be fused with other intelligence to give the commander a high level of situational awareness across the entire front. When an enemy target was confirmed, the commander could call for any direct or indirect weapon in range, including rifles, machine guns, grenade launchers, AT weapons, mortars, artillery, and air-delivered munitions. Because DWSC sensors could be distributed at significant depth, commanders at all levels could evaluate sensor input and respond by rapidly changing priorities of fire or locations of combat systems. This would allow a sequenced attack against numerous enemy targets by the most appropriate combat system.

The gateway could be programmed to interface with hand-emplaced or remotely delivered nonlethal weapons, AT mines, other sensors, or nearby outposts or patrols. If the situation indicated a need for remotely delivered AT mines, the gateway could also be programmed to communicate with and direct action of friendly AT mines in the vicinity. Through the gateway, the mines could be activated or deactivated by a man-in-the-loop in the operations center. These mines could also be part of a network that would communicate locations to each other, assess the situation based on programmed logic, and detonate according to programmed logic or human command.

The DWSC would not depend on the use of mines. It would be a sensor-based intelligence system that could be used in combination with different systems as the situation dictated, and would move warfare toward an integrated system of sensors, communications links, and combat systems. Because DWSC would exploit existing and future combat system capabilities and would not rely on a dedicated, unique kill mechanism, it might be more cost effective than some other mixed-system alternatives to APL. A concept of operations would have to be articulated throughout the user community, and doctrine for use would have to be developed prior to fielding. Because this system is still in the early concept phase, considerable research and development would be required to develop, produce, and field the DWSC.

Advantages

- Because the DWSC does not include APL, it would be conceptually compliant with the CCW Amended Protocol II and the Ottawa Convention.
- Rapid emplacement of sensors, gateways, AT mines (when required), and other systems could be valuable for protecting early-entry units or other friendly units.
- The man-in-the-loop capability should minimize friendly and noncombatant casualties out to significant depths on the battlefield.
- The DWSC would increase lethality by improving commanders' situational awareness and enabling them to bring combined arms capabilities to bear, as necessary.
- The DWSC would support emerging concepts of the future, nonlinear battlefield peopled by highly mobile, moderately armed forces.
- The system would provide a measure of security for AT mines, sensors, and other forward-deployed capabilities.
- The presence of DWSC could impart a psychological fear of death or injury throughout the battlefield.
- The presence of a self-destruct feature on any remotely delivered mines would address humanitarian concerns.
- The antihandling device on remotely delivered mines would have a psychological impact on dismounted enemy forces.

Disadvantages

- Because the system has no APL or human sensors, the mines could be very susceptible to clearance by a dismounted enemy, especially at some distance from friendly forces.
- Costs of development, production, and fielding would be very high. Given the complexity of the system, maintenance and logistics costs could also be high.

Raptor

Source: U.S. Department of Defense/Office of the Project Manager, Mines, Countermine, and Demolitions (Strano, 2000)

Raptor, a system being developed by DARPA, based on an Army science and technology program, has been described as a smart, autonomous, antiarmor/antivehicle system that provides situational awareness and targeting information to Hornet/WAM and other shooters. The system consists of an advanced overwatch sensor, deployed at depth, that can detect, track, and classify individual vehicles and recognize larger combat formations. The initial or "core" capability is based on hand-emplaced components delivered by truck or helicopter up to 25 kilometers from the forward line of friendly troops.

Vehicle-detection information from the advanced overwatch sensor would be electronically relayed through a gateway to "wake up" nearby Hornet/WAM AT munitions[2] in a stand-by, power-saving state. The gateway would be

[2] Raptor could also be implemented with other kill mechanisms, such as a mine patterned after BAT warhead technology.

programmed with the commander's tactics in selectable tactical "templates" to coordinate automatically the Hornet/WAM attack based on the threat. The system would be controlled by a Raptor control station located at the brigade tactical operations center.[3] The link between the gateway and the control station might be through radio line-of-sight transmission aided at extended ranges by antennas. The operator would automatically receive the exact location of the Raptor field components, including munitions, and would be able to turn the field on or off and select tactics remotely for the gateway.

A Raptor system could be modified to include a space-based, near real-time communication link so that joint tactics would be possible. With this improvement, the sensor target information would be relayed through the Raptor control station to the army tactical command and control system. The commander could then initiate additional checks by other sensors, such as unmanned aerial vehicles, and take action, such as directing the gateway to attack the threat by activating mines according to an emplacement template, targeting indirect-fire weapons, and/or alerting nearby combat units.

The fully developed Raptor system could be used for operations as much as 300 kilometers away with satellite and other communication links; mines deployed by powered parafoil or airdrop; mines embedded with the ability to discriminate between friend and foe; and sensors capable of detecting, reporting, and targeting light wheeled vehicles, low-flying aircraft, and artillery/missile firings. Incremental upgrades, including links to other Army tactical communications system, and other capabilities, should be ready for production at various times up to fiscal year 2008.

Raptor could move antivehicular mine warfare toward an integrated system of sensors, communication links, mines, and other combat systems. The concept of operations would be articulated throughout the user community, and doctrine would have to be developed prior to fielding. Because this system is still in the early concept phase, considerable resources would be required to develop, produce, and field Raptor.

Advantages

- Because there are no APL in this system, it appears to be conceptually compliant with the Ottawa Convention and CCW Amended Protocol II.
- The self-destruct feature would further address humanitarian concerns.
- The man-in-the-loop should minimize friendly and noncombatant casualties out to significant depths on the battlefield.
- Raptor would increase lethality by coordinating the attacks of two or more mines against detected vehicles, as determined by situation-dependent tactical templates.
- The system would dynamically integrate AT munitions into a combined joint sensor/killer team.
- The presence of Raptor would impart a fear of death or injury throughout the depth of the battlefield.
- The Hornet/WAM has an antihandling device to inhibit tampering, which may have a psychological impact.

Disadvantages

- The hand emplacement of vehicle sensors, gateways, and Hornet/WAM would dramatically increase manpower requirements and delivery time, especially when used at depth and across wide fronts. Emplacement by unmanned helicopters (if developed) or similar means could facilitate the use of Raptor in areas beyond the direct control of friendly forces.
- The absence of APL or human sensors could make Raptor mines susceptible to clearance by a dismounted enemy, especially at a distance from friendly forces.
- Development, production, and fielding costs would be high.
- Given the complexity of the system, maintenance and logistics costs could also be high.

Remote Antiarmor Mine System Enhanced with Telemetry and Sensor Package

Source: Committee on Alternative Technologies to Replace Antipersonnel Landmines

The Remote Antiarmor Mine System (RAAMS) enhanced with telemetry and a sensor package (RD-Sensor) would have the enhancements proposed in the RD-Telemetry (RAAMS enhanced with telemetry) concept. RD-Sensor would also contain a sensor package in addition to the AT mines. Once delivered, the AT mines and sensor packages would fall to the ground, deploying one or more low-power, miniature sensors. At least one of the sensors should be imaging (infrared, motion detection, video), with others being acoustic, seismic, or other types. The lowest power sensor, probably acoustic, would first detect vehicles or dismounted intruders and then "wake up" the more power-consuming imaging sensor.

A transmitter would send appropriate imagery information back to a friendly operations center, most likely located tens of kilometers away. If enemy dismounted forces were identified, the man-in-the-loop could call for indirect fire to keep the enemy from clearing the AT mines, reinforce the minefield with more AT mines, or take other action.

To reduce power and bandwidth requirements, one (or a limited number) of RD-Sensor projectiles/dispensers could be added to a minefield consisting of many remotely delivered, but not sensor-equipped, AT mines. RD-Sensor could also be incorporated into air-delivered or ground-delivered scatterable minefields.

[3] The committee questioned whether Raptor might be more appropriately controlled at a level lower than the brigade level.

Because RD-Sensor would provide near real-time knowledge of the location of remotely delivered minefields, the system could be used at the last minute against high payoff targets in support of mobile and other operations. This would reduce the time enemy forces would have to locate and clear remotely delivered AT mines. The integrated sensor package could also give the commander remote ears and eyes so he could call for indirect fire or take other actions to protect the AT mines from intruders. Near real-time sensor information would give the commander the option of reinforcing the minefield with additional remotely delivered mines or cover it with other AT weapons, such as SADARM, BAT, SFW, attack helicopters, or close air support while enemy vehicles were delayed by the AT mines. Antihandling devices on 20 percent of the munitions would discourage tampering.

RD-Sensor would require a considerable amount of research and development. To develop and integrate the sensor package into an existing dispenser, a number of technological challenges would have to be overcome: hardening and militarizing the sensors; developing enabling sensors that could determine their location and orientation; developing of infrared sensors that could "pop up" above vegetation and other obstacles; developing a limited, autonomous network of deployed sensors that could communicate with each other; devising methods of overcoming power and bandwidth challenges to allow nonline-of-sight communication of infrared imagery and other sensor feedback to friendly operations centers located many kilometers away.

Advantages

- Because RD-Sensor does not include APL, it would be conceptually compliant with CCW Amended Protocol II and the Ottawa Convention.
- The presence of a self-destruct feature and a man-in-the-loop would further address humanitarian concerns.
- Twenty percent of the mines would have antihandling devices to inhibit tampering.
- A more precise estimate of the ground location of remotely delivered mines might reduce fratricide.
- More certain knowledge of the location of remotely delivered mines in critical locations would enable combat commanders to use remotely delivered AT mines at the last minute, thus reducing an enemy's ability to find and clear the minefield.
- With information on an attempted breach of an AT minefield or certainty that enemy vehicles are being destroyed, a commander could take action to protect the AT mines or reinforce this success.

Disadvantages

- A significant number of delivery assets would be required to emplace a large minefield.
- On current mines, self-destruct times can not be reset from the factory-set times.
- Current mines do not have a command-destruct feature.
- Research, development, and acquisition costs would be high.

Remotely Delivered Hornet/Wide Area Munition

Source: Committee on Alternative Technologies to Replace Antipersonnel Landmines

The Hornet/WAM is a hand-emplaced, autonomous AT mine currently entering inventory. The proposed alternative (RD-WAM) would add a remote delivery capability, via a Volcano launcher or from a deep-attack asset, such as MLRS or TACMS. It might also be possible to emplace the mine via gravity ordnance (such as Gator). The Hornet/WAM weighs about 16 kilograms, is about 36 centimeters high, and about 23 centimeters in diameter. When its seismic sensors detect movement, it alerts the mine to turn on its acoustic sensors, which detect and classify the target. If an armored target approaches within 100 meters, a submunition with an infrared sensor is launched over the target and fires an explosively formed projectile into the engine compartment. Hornet/WAM AT mines have an antihandling feature that causes the mine to detonate when disturbed. The mine is designed to operate for 30 days and then self-destruct.

The tactical use of a RD-WAM would have to be developed and added to doctrinal manuals. Although a comprehensive and lengthy research and development effort would be required to harden and otherwise modify the Hornet/WAM for remote delivery, other nations have developed hardened mines similar in size to Hornet/WAM. Thus, hardening of Hornet/WAM should be feasible.

Advantages

- RD-WAM would comply with CCW Amended Protocol II and the Ottawa Convention.
- The self-destruct feature would further address humanitarian concerns.
- The antihandling device would inhibit tampering.
- The RD-WAM provides wide-area coverage and an off-route capability.
- Remote delivery would increase the utility and range of use of Hornet/WAM on the battlefield.

Disadvantages

- Development and procurement costs would be high.
- Remote delivery would require the use of high-value assets.

Self-Healing Minefield

Source: DARPA Track II (Altshuler, 1999)

The self-healing minefield would be a dynamic,

scatterable AT minefield (the munition would be similar in size and delivery method to Volcano or Gator mines). Through mine-to-mine communication and interaction, individual mines would respond to breaching attempts by reorganizing (physically jumping) to fill in open lanes, thereby establishing a barrier. Because a breach could not be sustained, the enemy would be forced to change tactics from breaching to clearing the minefield. Thus, this system would be an alternative to mixed mine systems.

Before this technology can be transferred from DARPA to the Army, it will have to undergo several stages of testing: preliminary analysis to determine the validity of the concept; verification of its battlefield utility; development of enabling technologies; and testing of the technologies.

A preliminary modeling exercise has been completed by the Institute for Defense Analyses to demonstrate concept viability. Battlefield utility has been explored at Lawrence Livermore National Laboratory through a simulation of a single scenario, which had favorable results. According to the program manager, Sandia National Laboratories have demonstrated preliminary physical/mechanical capability (a 12-centimeter diameter, 2.27-kg payload and actuator using liquid fuel was shown to travel 6 to 7.5 meters in the air). According to the program manager at DARPA, contracts of up to three years have been concluded for all aspects of the program. At the conclusion of these contracts the viability of the mine and technology for a 50-mine minefield should have been demonstrated. (Telephone conversation between Dr. Altshuler, DARPA Program Manager for Antipersonnel Landmine Alternatives, and Study Director about status of self-healing minefield progress, August 14, 2000.) The mobility target would include the capability of a mine to perform multiple jumps over hundreds of meters; the durability would be comparable to Volcano.

The developmental issues that must still be addressed can be categorized as mechanical issues and communications-related issues. Mechanical issues include mine mobility and distribution. The ability of a mine to jump distances of a few meters several times over will have to be convincingly and repeatedly demonstrated, as will the ability of a mine to jump in wooded areas, shrubs, and muddy terrain. Predetermined distribution (geolocation) of mines, particularly scatterable mines delivered by aircraft (fixed-wing or rotary-wing), will have to be demonstrated or shown to be unnecessary. The current DARPA program includes limited use of GPS; therefore system viability during periods when GPS is denied will have to be demonstrated. The response time to a breach will have to be ascertained and shown to be tactically significant.

According to the DARPA program manager, mine-to-mine communication technology and reliability are not considered major developmental issues because technology that is almost "off the shelf" can be used. The issue of vulnerability to countermeasures (jamming and spoofing) will have to be addressed, as well as whether it is possible to develop an advanced warhead with the same diameter and volume, but with increased penetration capability. Depending on the capabilities of the mines, the mine density in a field might have to be changed, which might require changes in doctrine or tactics, techniques, and logistics.

Advantages

- Because these are AT mines, they are compliant with the Ottawa Convention and would be configured to comply with the CCW Amended Protocol II.
- A self-healing minefield, although a long-term technology, would use advanced technologies and incorporate innovative, out-of-the-box thinking.

Disadvantages

- The projected cost of laying a minefield is anticipated to be about twice that of laying a Volcano field.
- It is likely to take a long time to bring this item into production.

BAT Antiarmor Munition

Source: Committee on Alternative Technologies to Replace Antipersonnel Landmines

During the late 1960s and early 1970s, the Army developed to the prototype stage a distributed-sensor, antiarmor mine known as HOMINE (for homing mine), which consisted of a sensor field covering an area similar to that of an AT minefield and a separately located kill system. The BAT Antiarmor Munition (BATAAM) resurrects the basic HOMINE concept and broadens its capability by using a BAT as the kill system with the submunition and a "minefield" surveillance system controlled by a remote operator.

The BAT submunition (described in Chapter 5) was originally designed to be delivered by missile or aircraft, at long range, for a many-on-many attack against armored columns. In the BATAAM concept, the same munition would be positioned on the ground and activated by sensors. The BATAAM concept would lend itself to two deployment options.

Option 1. BATAAM could eventually be a replacement for WAM PIP, the advantage being its broader kill radius. Equipped with the WAM PIP detection/localization sensors (acoustic/seismic), and incorporating a GPS receiver and antihandling device, the single-weapon BATAAM launch canisters would be hand emplaced in the centers of the BAT target acquisition footprint (the spacing between weapons would be an abutting grid of such footprints). The minefield surveillance package would consist of infrared/optical sensors, communication equipment, and a GPS receiver linking the package to both the individual weapons and an operator. Surveillance-to-operator communications would be line-of-sight, radio-frequency transmissions for short distances (3 to 5 kilometers) and aircraft or satellite-link communications for longer transmissions. Each BAT sensor would remain on to alert the surveillance system but could only be fired when enabled by the operator (after which it could be activated by its own sensors). In addition, the operator would have the ability to fire or command detonate each weapon. A timed, self-deactivation feature and antihandling device would provide backup in the event the operator were not able to perform this function.

Option 2. In the second deployment option, the minefield would have small, hand-strewn or helicopter-deployed magnetic/pressure sensors. The surveillance/communications package and the multiple-shot BAT kill system could be colocated or separately positioned on the periphery of the minefield. A radio-frequency signal from a small sensor would wake up the surveillance system. With the locations of the surveillance and kill system determined by the GPS, and using a topographic map, the operator could superimpose a grid on the minefield if necessary. Individual BAT weapons, equipped with propulsion to loft them from the canister to an altitude for target surveillance within the minefield, would be command-fired by the operator.

As in Option 1, the surveillance and kill system would be equipped with an antihandling device, but in Option 2, the weapon could only be fired by the operator, thus allowing for friendly transit as well as optimum enemy force deployment within the minefield. The operator would have the option of command detonating the BAT weapons. In both options, the weapon system would be equipped for self-deactivation in the event of operator neutralization.

The WAM PIP sensors and the BAT upgrade are under development. The components of the surveillance package are available, but would have to be assembled and militarized. Tactical employment of the BATAAM would have to be added to doctrinal publications as the system is fielded.

Advantages

- This AT mine would comply with the Ottawa Convention and could be made to comply with CCW Amended Protocol II.
- BATAAM would provide for a less expensive means of delivering the BAT submunition than the present missiles.
- Weapon delivery against targets would be more precise.
- An AT minefield could be established for close-in protection (battlefield conditions permitting). In Option 2, the sensor could be delivered by missile, artillery, or aircraft.
- BATAAM would allow safe passage by friendly forces.
- Both deployment options would permit optimum deployment of intruding forces before weapon release.
- Expensive components could be command detonated or retrieved for reuse.

Disadvantages

- BATAAM must be hand emplaced; in high-visibility terrain, BAT canisters might require burial.
- The system would be somewhat more expensive than other alternatives.
- BATAAM would require a man-in-the-loop for optimum effectiveness.

Early-Warning Subsystem for Remotely Delivered Antitank Minefields

Source: Committee on Alternative Technologies to Replace Antipersonnel Landmines

The Early-Warning Subsystem (EWSS) for remotely delivered AT minefields would be based on small electronic components integrated into Volcano, Gator, RAAMS, and other scatterable AT systems to transmit their location, direction, and relative position. Friendly forces would be equipped with receivers that could interpret subsystem transmissions to provide real-time information on nearby, friendly, scattered minefields. With this situational awareness, friendly forces could maneuver around the minefields, thus reducing the likelihood of fratricide.

With this system, next-generation AT mines might be programmed with additional self-destruct times. Very brief self-destruct times, perhaps fractions of an hour, would have several advantages. First, a commander would have the option of using an extremely short-duration, precise minefield that would support the plan of maneuver and minimize fratricide and the risk to noncombatants after friendly forces had passed. Second, it would enable commanders to use scatterable minefields in high-risk locations in all weather conditions, just in time, and close to friendly troops. Much longer self-destruct times of up to 30 days or more could be used for static situations where there were few noncombatants.

Ideally, transmissions from the subsystem could be received by and displayed on the screens of standard mounted and dismounted digital command and control devices. As an interim alternative, a small, single-purpose receiving device could be developed to digitally display warning signals from scatterable minefields. The receiver could also include a command-destruct option to allow AT mines to be detonated as the situation warranted.

Tactics, techniques, procedures, and training devices would have to be developed to teach soldiers to interpret and react to warning signals emanating from actual or simulated remotely delivered minefields. Emulators could be routinely used during field training exercises to increase the organization-wide understanding of and confidence in using scatterable AT mines in combat. With the high level of situational awareness, scatterable minefields could be emplaced at the last possible minute, thereby providing AT mines with a measure of protection by giving the enemy less time to react and breach the minefield.

Considerable research and development assets would be required to develop the transmitter, receiver, and other hardware/software necessary to implement this concept. User documentation would also have to be developed to articulate the battlefield need for an EWSS.

Advantages

- The EWSS would use AT mines that are compliant with the CCW Amended Protocol II and the Ottawa Convention.
- An EWSS would improve the situational awareness of friendly units and reduce the likelihood of fratricide.
- Precise information about the location of scatterable minefields could lead to shorter self-destruct times for AT mines.
- A command-destruct option would be viewed positively by friendly forces and humanitarian organizations.
- The use of early-warning emulators in training could increase the confidence of commanders and soldiers in using scatterable minefields in combat.

Disadvantages

- Information provided by EWSS would have to be integrated into digital displays.
- The minefields could be easily breached if the enemy obtained the information provided to friendly forces.
- Research and development costs could be high.

Remote Antiarmor Mine System with Nonlethal Capability

Source: Committee on Alternative Technologies to Replace Antipersonnel Landmines

A modification of the RAAMS concept, the Remote Antiarmor Mine System with nonlethal capability (RAAMS-NL) would include eight current RAAMS AT mines augmented with Taser nonlethal munitions in a 155-mm howitzer projectile. The purpose of the Taser nonlethal munitions would be to provide protection against a dismounted enemy breaching force. When the infrared sensors detect a human intruder, the munition would propel small barbed darts out to 6 meters. When the darts entered an intruder's skin or clothing, an incapacitating electric shock of 50,000 volts would be produced in 4 to 6 microsecond pulses, 10 to 20 times per second. The current Taser nonlethal munition power supply could support approximately 10 minutes of continuous operations. However, the electric shocks could be cycled less frequently to incapacitate an enemy for several hours.

Existing doctrine and tactics should be adequate for this system, although publications and firing tables would have to be adjusted. Gun-hardening, modifying, and integrating the Taser nonlethal munitions into the RAAMS-NL projectile would require significant research and development. Development of a long-term, reliable power supply could present a major challenge.

Advantages

- The AT mines and Taser nonlethal munitions comply with the CCW Amended Protocol II.
- Even though the Taser nonlethal munition fires electrically charged darts, the electric shocks are considered to be nonlethal, which should make this system acceptable under the Ottawa Convention.
- Taser nonlethal munitions would provide the AT minefields a measure of protection against dismounted breaching forces.

Disadvantages

- The effects of Taser nonlethal munitions on humans (particularly children) are not fully known. However, police forces that use Taser devices have collected large amounts of data in this area.
- The Taser is not lethal and, therefore, would have limited psychological and physical impacts against a determined enemy.
- Intruders could deflect darts by carrying antiriot shields.
- Battery/power issues would increase maintenance and storage costs over the life cycle of the system.

COMMITTEE ASSESSMENTS

Materiel Alternatives Against Dismounted Targets

The committee considered five systems for use against dismounted targets that should or could be available after 2006. Compared to the M14/M16 baseline system, all of the systems appear to meet both the military and humanitarian requirements developed by the committee. Several of the alternatives would provide graduated responses and could be used either in pure APL modes or as mixed munitions. They are presented below beginning with the alternative that scored highest in military criteria, but the committee believes all of them warrant consideration.

The RRASMS would use an electronically programmable radio to facilitate communications and self-locate. The sensor information would be provided to a soldier/operator who could select a response from a menu of increasingly lethal modular munitions.

The URAS would use Claymores appropriate to the nature of the terrain and the anticipated size and distribution of intruding forces. URAS would provide firing versatility for maximum effectiveness.

A DARPA Track II concept, the Tags/Minimally Guided Munitions, could detect and locate dismounted enemies through tags that affix themselves to enemy soldiers. The tag-cued munition, released after a man-in-the-loop decision, would make possible post-apogee trajectory changes to focus on the target (tag) rather than an area.

The LDMG would use laser radar and an automatically

aimed machine gun. The soldier/operator would decide whether the response should be lethal or nonlethal.

A distributed-sensor AP minefield could be used as an area-denial weapon or in mixed mode as protection for an AT minefield. The kill system would have to be installed by hand. The system would rely on remote observations by a man-in-the-loop who had been alerted by optical and infrared sensors.

Materiel Alternatives Against Mounted Targets

After 2006, it will be technologically feasible to improve the tactical effectiveness of existing or proposed remotely delivered AT mines by providing multiple means of remote deployment and additional resistance to countermeasures through signature reduction and the use of nonlethal techniques. The committee considered eight systems that might be available after 2006 that could be used against mounted enemies. Compared to the Volcano baseline system, all of them appeared to meet the military and humanitarian requirements for an APL alternative (see Table 7-2 for score sheet).

Mixed Antitank/Nonlethal Alternatives

RAAMS-NL would modify the existing RAAMS munition to include a Taser nonlethal device activated in some manner (e.g., by a trip wire). The Taser would protect the deployed AT mines from attempts to breach them by subjecting an intruder to high-intensity electric shocks. Although gun-hardening, modifying, and integrating the Taser munitions into the RAAMS projectile would require significant research and development, the RAAMS system would be greatly improved by the addition of a nonlethal antipersonnel component.

Pure Antitank Mine Systems

The Raptor, already in development, would be a smart, autonomous, AT system that would improve situational awareness and provide targeting information for other weapons, such as Hornet/WAM. The system would be deployed, initially hand-emplaced, deep in the battlespace to detect, track, and classify individual intruding vehicles. Soldier/operators at the brigade tactical operations center would know the exact locations of the Raptor components/munitions and would be able to operate these components from a distance. Although Raptor would be very expensive, the system could be highly effective, especially with remote delivery modifications. Raptor received high scores in the military-effects category.

RAAMS enhanced with telemetry and sensor package (RD-Sensor) would add a sensor package to the RD-Telemetry alternative described in Chapter 6. Following delivery, the AT mines and sensor package would deploy one or more low-power, miniature sensors to detect vehicles or dismounted intruders. A transmitter would send appropriate imagery information back to the soldier/operator who could determine the appropriate course of action.

Recommendation. The Army should proceed rapidly with plans for modernizing existing remotely delivered pure antitank landmine systems, such as RAAMS and Volcano (M87A1), by incorporating other technologies, including sensors, precision locators, and nonlethal devices.

The remotely delivered RD-WAM would be an enhancement to the current Hornet/WAM, which requires hand emplacement. A modified, hardened Hornet/WAM could be remotely delivered using a Volcano launcher, an MLRS, or a tactical missile system. The remotely delivered Hornet/WAM received higher combined scores in the areas of technological risk and cost than did other alternatives. The committee believes this alternative has great potential because the Hornet/WAM is a superior weapon except for its limited means of delivery.

The Self-Healing Minefield, a DARPA Track II program, would be an intelligent, distributed network of AT mines with decentralized control. Unlike many of the other alternatives, the Self-Healing Minefield would not have a man-in-the-loop. Individual munitions would detect breaching attempts through mine-to-mine communications and automatically react by moving to fill gaps in the minefield. This promising innovative system is unlikely to be available for at least 10 years.

Recommendation. The development of the Self-Healing Minefield concept, which automatically reacts to any breaching attempt by refilling gaps, should be experimentally evaluated to determine its operational effectiveness.

The BATAAM provides a means of using the smart munition BAT in a static minefield situation. Although it provides no increased military effectiveness over current systems, the introduction of the sensor field provides the commander with greater battlefield flexibility.

Sensor Systems

The EWSS for remotely delivered AT minefields would integrate small electronic components into scatterable AT systems to transmit their location to friendly forces equipped with receivers who would then be able to maneuver around minefields. EWSS would reduce fratricide and allow more confident use of the "just-in-time" delivery of AT mines to discourage breaches and reinforce high-risk locations. The committee believes that EWSS could leverage emerging information systems technologies.

DWSC would be a sensor network linked to existing combat systems by a communications network. DWSC would involve delivering, by artillery or air, hundreds, or even thousands, of small, expendable sensors over a wide area that would communicate data back to a central point. Using a

TABLE 7-2 Score Sheet for Alternatives Potentially Available After 2006

		System Name	RRASMS	URAS	Tags	LDMG	Distri-Sensor Minefield	DWSC	Raptor	RD Sensor	RD-WAM	Self-Healing Minefield	BATAAM	EWSS	RAAMS-NL
		Available for Implementation	>06	>06	>06	>06	>06	>06	>06	>06	>06	>06	>06	>06	>06
		Baseline D=M14/16 M=M87	D	D	D	D	D	M	M	M	M	M	M	M	M
	A0	**Military Effects** Sum:	**15**	**12**	**10**	**7**	**5**	**14**	**12**	**4**	**3**	**1**	**0**	**–1**	**–2**
Mounted	A1	Enhances effects of close and deep friendly fires						3	3	1	3	1	1	0	–1
Mounted	A2	Has multiple methods of delivery						0	–1	0	0	0	–1	0	0
Mounted	A3	Provides a range of effects that inhibit mounted and dismounted maneuvers						2	2	1	–1	–1	–1	–1	–1
Mounted	A4	Resists full spectrum of enemy breach methods, including dismounted methods						3	3	1	2	3	–1	–1	–1
Mounted	A5	Provides early warning of ground attack						2	2	1	0	0	0	0	0
Mounted	A6	Is safe for friendly forces						0	0	0	0	0	0	0	0
Mounted	A7	Is effective in all types of terrain and weather						–1	–1	–1	–1	–1	0	0	0
Mounted	A8	Poses minimal residual hazard to own forces and noncombatants after military conflicts						1	1	0	0	0	1	0	0
Mounted	A9	Is difficult to detect by enemy forces						–1	–1	–1	–1	–1	–1	–1	0
Mounted	A10	Poses minimal risk of fratricide						3	3	2	1	1	1	3	1
Mounted	A11	Has modifiable effects for mounted and/or dismounted threat						2	2	1	0	0	0	0	0
Mounted	A12	Has controllable activation/deactivation and duration before and after installation						1	2	0	0	0	2	0	0
Mounted	A13	Is effective in nuclear, chemical, and biological environments						–1	–1	–1	0	–1	0	–1	0
Mounted	A14	Is easy and efficient to distribute						0	–2	0	0	0	–1	0	0
Dismounted	A15	Can delay, disrupt, and/or canalize enemy movement/maneuvers	2	2	2	1	1								
Dismounted	A16	Denies enemy access to terrain/facilities (including short- and long-term deterrent for boundaries and DMZ areas)	2	2	2	1	1								
Dismounted	A17	Enhances effects of friendly-force weapons, obstacles, and munitions (including AT mines)	2	2	2	2	2								
Dismounted	A18	Generates exploitable delays and opportunities (fixes or contains enemy)	2	2	2	1	1								
Dismounted	A19	Generates detection, alert, and/or early warnings	2	2	1	–1	–1								
Dismounted	A20	Facilitates classification of target	2	2	0	1	1								
Dismounted	A21	Produces desired effects on enemy forces (nonlethal to lethal)	2	0	0	2	0								
Dismounted	A22	Reduces casualties/risk for U.S. and/or allied forces	2	–1	2	1	1								
Dismounted	A23	Deters pursuit to facilitate breaking of contact under pressure	–1	1	–1	–1	–1								
	B0	**Humanitarian Concerns** Sum:	**4**	**3**	**5**	**5**	**5**	**6**	**6**	**4**	**4**	**2**	**3**	**2**	**2**
	B1	Reduces postconflict residual hazard (CCW)	1	1	1	1	1	1	1	1	1	1	1	1	1
	B2	Location can be recorded (CCW)	1	1	1	1	1	2	2	2	0	0	1	0	0
	B3	Detectable to facilitate removal (CCW)	0	1	1	1	1	1	0	0	0	0	0	0	0
	B4	Discriminates between combatants and civilians (Ottawa)	1	2	1	1	1	1	2	0	2	0	0	0	0
	B5	Does not explode on presence, proximity, or contact of a person (Ottawa)	1	1	1	1	1	1	1	1	1	1	1	1	1
	C0	**Overall Technical Risk (C4ISR, etc.)**	**–3**	**–2**	**–3**	**–3**	**–3**	**–3**	**–2**	**–3**	**–2**	**–2**	**–2**	**–2**	**–3**
	D0	**Requires Change in Tactics/Doctrine**	**Y**	**Y**	**Y**	**Y**	**Y**	**Y**	**Y**	**Y**	**Y**	**Y**	**Y**	**Y**	**Y**
	E0	**Cost** Sum:	**–6**	**–5**	**–6**	**–6**	**–6**	**–6**	**–6**	**–6**	**–5**	**–6**	**–4**	**–6**	**–6**
	E1	Research and development	–3	–2	–3	–3	–3	–3	–3	–3	–2	–3	–2	–3	–3
	E2	Procurement	–3	–3	–3	–3	–3	–3	–3	–3	–3	–3	–2	–3	–3

NOTE: Acronyms are defined on p. xv in the front matter.
Key: D = dismounted, M = mounted, n/a = nonapplicable.

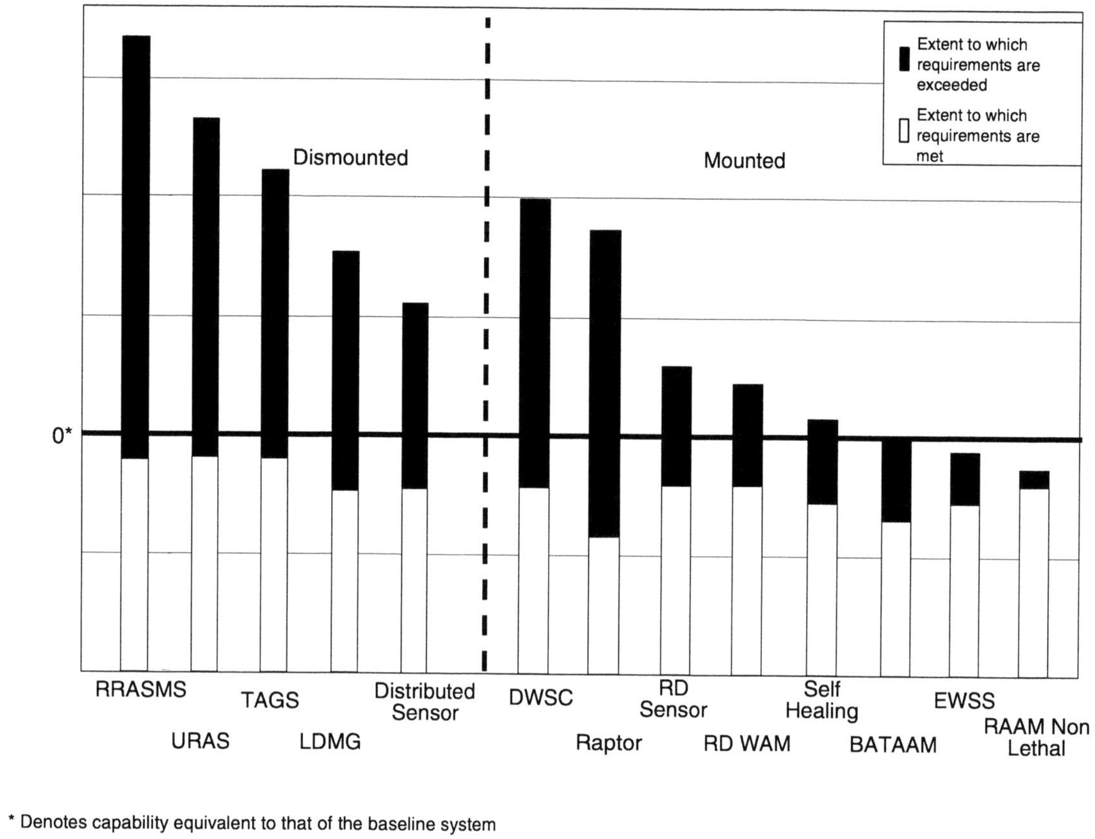

FIGURE 7-1 Military effectiveness of alternatives potentially available after 2006 based on qualitative scoring by the committee.

man-in-the-loop, a commander could make an informed decision about an appropriate response to an intrusion. DWSC would exploit existing and developing communications and combat systems. Because DWSC would not require a dedicated, unique kill system, it could be more cost effective than some other APL alternatives. This integrated approach would be one of the most effective future systems and scored very high in the military effects category. In addition, DWSC is already the focus of a major Army science and technology program. However, several technology issues would have to be solved before the system could become operational.

Summary

The criteria and scores are displayed in tabular form in Table 7-2. The details of how these scores were derived can be found in Chapter 4.

Figure 7-3 is a graphical summary of the scoring. In keeping with the Statement of Task, this graph shows only the relative military effectiveness of candidate systems without regard to cost, risk, or humanitarian factors. Each bar on the graph is a composite. The lower portion (white) shows the degree to which each system meets the military effectiveness requirements in comparison to the baseline system. If the candidate system meets all of the requirements at least as well as the baseline system, the score is 0. If it is less effective in any requirement, the score is less than 0. The upper portion (dark shading) of the bar shows capabilities that exceed those of the baseline system.

These graphs use the methodology described in Chapter 5. In general, if the total bar height is high, the system is likely to be militarily effective. If the value of the lower portion of the bar is near 0, the system meets most of the military requirements. If the lower bar is much lower than 0, the system probably has significant differences from the baseline mine and will not perform some desired functions. However, that system may still be militarily effective if it performs some functions much better than the baseline system. Because the scoring criteria were not weighted, these graphs should be used only for assessing trends and making qualitative comparisons.

8

Conclusions and Recommendations

President John F. Kennedy closed his address to the Centennial Convocation of the National Academy of Sciences quoting French Marshall Lyautey, who when talking with his gardener said, "Plant a tree tomorrow." The gardener replied, "It won't bear fruit for a hundred years." "In that case," said Lyautey, "plant it this afternoon." (News Report, 1963)

INTRODUCTION

Conclusion 1. The major reasons for seeking alternatives to current antipersonnel landmines (APL) are humanitarian concerns, compliance with the Ottawa Convention, and enhanced military effectiveness. Indeed, this study would not have been empanelled were it not for the Ottawa Convention. The current inventory of self-destructing and self-deactivating U.S. APL is militarily advantageous and safe. They achieve desired military objectives without endangering U.S. warfighters or noncombatants more than other weapons of war, but they are not compliant with the Ottawa Convention. However, humanitarian concerns and Ottawa compliance are not always synonymous. In fact, some of the apparently Ottawa-compliant alternatives examined by the committee may be less humane than present U.S. self-destructing and self-deactivating landmines.

Recommendation 1a. If the decision is made to accede to the Ottawa Convention, a transition period may be necessary before implementation to maintain current U.S. military capabilities until suitable alternatives can be made available. During that transition, existing self-destructing and self-deactivating antipersonnel landmines should be retained, both in their stand-alone form and as part of mixed systems.

Recommendation 1b. Of the solutions not compliant with the Ottawa Convention, simply retaining the current self-destructing and self-deactivating mines would be the best course of action.

Conclusion 2. The rapid emergence of new technologies after 2006 will create opportunities for the development of systems that can outperform today's antipersonnel landmines and that would be compliant with Ottawa.

Recommendation 2a. The development of sensor-net technology should be pursued aggressively and eventually incorporated into a fully militarized, deployed system characterized by networking, strong detection and tracking capabilities, robustness, low power consumption, low cost, covertness, low probability of intercept, easy deployment, and disposability.

Recommendation 2b. Investments already being made in new technologies for other purposes should be leveraged and applied to the search for alternatives to antipersonnel landmines.

ALTERNATIVES AVAILABLE BY 2006

Conclusion 3. By 2006, alternative tactics or operational concepts could not, on their own, provide tactical advantages similar to those provided by antipersonnel landmines, without a significant increase in force structure. In certain situations, however, some nonmateriel alternatives might be useful: increased reconnaissance forward; more soldiers or weapon systems in a given battlefield area; more command-detonated Claymores to protect against a dismounted enemy; antitank mines remotely delivered "just in time" to support a maneuver and inhibit the enemy's ability to breach; and speed, mobility, and offensive tactical operations.

Conclusion 4. For use against dismounted forces, the Track I alternative to nonself-destructing landmines (NSD-A) could provide, by 2006, similar or enhanced tactical advantages for U.S. forces as compared to those provided by current nonself-destructing antipersonnel landmines. The battlefield override switch, a software capability that allows the system to operate autonomously, is highly contentious because, as presently designed, it would render the NSD-A non-Ottawa compliant. Even though the timing of a decision on the switch or other programmatic delays could jeopardize the timeline, the NSD-A system appears to be technically mature enough to be available by 2006. This weapon system could be greatly enhanced in the future by planning for the inclusion of additional sensors, nonlethal elements, and an Ottawa-compliant battlefield override capability.

Recommendation 4a. The development and production of the Track I alternative to nonself-destructing landmines (NSD-A) system should be aggressively pursued to ensure its availability by 2006.

Recommendation 4b. Two suites of weapon software should be developed simultaneously in preparation for a presidential decision concerning the Ottawa Convention. If compliance with the Ottawa Convention were desired, the battlefield override switch, as currently designed, would not be used in the production of the NSD-A. If the president decides that other considerations outweigh Ottawa compliance, the option of retaining the switch would be available. In any case, Ottawa-compliant variations to the battlefield override switch should be explored to provide the United States with greater flexibility.

Recommendation 4c. Sensor technology should be leveraged immediately to develop sensor systems to improve a soldier's ability to discriminate among friends, foes, and noncombatants in all terrain and all weather conditions at much greater battlefield ranges.

Conclusion 5. Under current policy, no fully equivalent alternative to mixed systems is likely to be available by 2006. Other than the Track III search for an alternative, little is being done that could lead to the fielding of a satisfactory alternative. The Hornet/Wide Area Munition (WAM), with its large lethal radius and antihandling device, could replace most of the tactical functions currently provided by mixed systems but has no remote delivery capability. If a satisfactory remote delivery capability could be developed by 2006, the Hornet / WAM appears capable of performing the mixed-minefield mission satisfactorily.

Recommendation 5a. Promising Track III concepts should be developed into weapon system programs. The development of any of these concepts by the 2006 deadline, however, would require that considerable additional resources be allocated for development and procurement.

Recommendation 5b. The feasibility, cost, and schedule of providing a remote delivery option for the Hornet/Wide Area Munition should be investigated. Shock hardening of the mine to withstand the impact of remote delivery appears to be an Ottawa-compliant, low-risk solution to current mixed minefields.

Conclusion 6. The Remote Area-Denial Artillery Munition (RADAM), a mixed system, provides little or no military advantage over the combined use of the Remote Antiarmor Mine System (RAAMS) and the Area-Denial Artillery Munition (ADAM). Because RADAM would be no more compliant with the Ottawa Convention than the ADAM/RAAMS combination, funding for its development could be better spent on accelerating the development of an Ottawa-compliant alternative. If DOD determines that an artillery-delivered mixed system must be maintained, there are two options: (1) request a change in presidential policy to allow the continued use of ADAM to be fired in tandem with RAAMS; or (2) develop RADAM. The latter option would require taking the Ottawa-compliant RAAMS out of the inventory to create a new non-compliant munition.

Recommendation 6. Until a long-term solution can be developed, the Area-Denial Artillery Munition (ADAM) should be retained in the inventory for use with the Remote Antiarmor Mine System (RAAMS). Production of the Remote Area-Denial Artillery Munition (RADAM) should be halted and funding redirected toward the development of long-term alternatives for mixed systems.

Conclusion 7. Although nonlethal variants by themselves cannot replace antipersonnel landmines, they would be useful in certain military operations. U.S. forces will face a broad range of potential scenarios in the future, from peace operations to intense full combat. With nonlethal variants, U.S. forces could mount a graduated response in situations where the threat is unclear, such as peace operations, or if large noncombatant populations were in the immediate tactical area. Nonlethal weapons have several advantages: they can be used in a broad variety of circumstances; they can be triggered automatically; and they do not require man-in-the-loop operation to be Ottawa compliant, which could improve the timeliness of a response and lessen the burden on the soldier/operator.

Recommendation 7. The development of nonlethal variants to support antipersonnel landmine alternatives should be emphasized. Funding should be restored and development accelerated for the nonlethal Canister-Launched Area-Denial System (CLADS). The CLADS munition should then be integrated into Volcano (M87A1) canisters to provide a mix of antitank and nonlethal antipersonnel munitions.

ALTERNATIVES POTENTIALLY AVAILABLE AFTER 2006

Conclusion 8. After 2006, improvements in the tactical effectiveness of existing or proposed remotely delivered antitank (AT) landmines ought to be technologically feasible, which could eliminate the need for mixed systems. Future systems that separate the sensor from the shooter could be improved by multiple means of remote deployment and resistance to countermeasures through signature reduction and other techniques. Track III programs, like the Track I initiative, will require concentrated effort and stable funding. In the long term, the emergence of new technologies, such as the ability to distinguish accurately between combatants and noncombatants, will provide opportunities for the development of systems that can outperform today's antipersonnel landmines.

Recommendation 8a. The Army should proceed rapidly with plans for modernizing existing remotely delivered pure antitank landmine systems, such as Remote Antiarmor Mine System (RAAMS) and Volcano (M87A1), by incorporating other technologies, including sensors, precision locators, and nonlethal devices.

Recommendation 8b. The development of the Self-Healing Minefield concept, which automatically reacts to any breaching attempt by refilling gaps, should be experimentally evaluated to determine its operational effectiveness.

Recommendation 8c. Several other technologies or systems already under development for other purposes should be considered as potential components of long-term alternatives to antipersonnel landmines, including unmanned air and ground vehicles, directed-energy weapons, battlefield sensory-illusion devices, passive transponders (e.g., tags), and other lethal and nonlethal systems.

SELF-DESTRUCTING, SELF-DEACTIVATING FUZES

Conclusion 9. The self-destructing and self-deactivating capability of today's U.S. scatterable landmines, used in accordance with international law, is a desirable operational capability because it (1) increases maneuver options and (2) addresses humanitarian concerns by reducing residual explosive hazards.

Recommendation 9. Any nonrecoverable, explosive alternative to antipersonnel landmines should have self-destructing and self-deactivating fuzes to meet operational requirements, address humanitarian concerns, and reduce fratricide among friendly troops. The U.S. government should consider equipping all nonrecoverable explosive munitions with similar technologies.

References

Altshuler, T.W. 1999. Defense Advanced Research Projects Agency (DARPA) Track II Antipersonnel Landmine Alternatives. Briefing by T. Altshuler, DARPA Program Manager for Antipersonnel Landmine Alternatives, to the Committee on Alternative Technologies to Replace Antipersonnel Landmines, National Research Council, Washington, D.C., December 10, 1999.

Berger, S.R. 1998. Letter from S.R. Berger, Assistant to the President for National Security Affairs, to Sen. Patrick Leahy (D-Vt) May 15, 1998.

Biering, M. 1999. U.S. Military Use of Anti-Personnel Landmines. Briefing by M. Biering, Engineer Team Chief, Office of the Deputy Chief of Staff for Operations and Plans–Force Development, Headquarters, Department of the Army, to the Committee on Alternative Technologies to Replace Antipersonnel Landmines, National Research Council, Washington, D.C., December 9, 1999.

Bornhoft, G. 1999. Strategic and Tactical Landmine Usage Overview. Briefing by G. Bornhoft, BRTRC Technical Research Corporation, to the Committee on Alternative Technologies to Replace Antipersonnel Landmines, National Research Council, Washington, D.C., November 15, 1999.

Braham, R. 1999. Technology 1999 Analysis and Forecast: Aerospace and Military. IEEE Spectrum 36: 73–78.

Carapezza, E. 2000. Tactical Sensor Program. Briefing by E. Carapezza, Defense Advanced Research Projects Agency (DARPA) Program Manager, Tactical Sensor Program, to a subcommittee of the Committee on Alternative Technologies to Replace Antipersonnel Landmines, Arlington, Virginia, March 13, 2000.

CJCS (Chairman of the Joint Chiefs of Staff). 1996. Joint Vision 2010. Washington, D.C.: U.S. Department of Defense, Office of the Chairman of the Joint Chiefs of Staff.

CJCS. 1997. Concept for Future Joint Operations. Washington, D.C.: U.S. Department of Defense, Office of the Chairman of the Joint Chiefs of Staff.

CJCS. 2000. Joint Vision 2020. Washington, D.C.: U.S. Department of Defense, Office of the Chairman of the Joint Chiefs of Staff.

Clark, W.K. 1998. Testimony before the House National Security Committee by W.K. Clark, Commander in Chief, United States European Command, Congressional Record. 1998 105th Congress. 2nd Sess. Vol. 144, No. 21. Available on line: *http://www.house.gov/hasc/testimony/10th5congress/3-5-98clark.htm.*

Clinton, W.J. 1997. Remarks by President Clinton on Land Mines. Available on line: *http://www.whitehouse.gov/WH/New/html/19970917-8619.html.*

Close, G.F. 1999. Joint Vision 2010. Briefing by G. Close, Director for Operational Plans and Interoperability–J-7, Office of the Joint Chiefs of Staff, to the Committee on Alternative Technologies to Replace Anti-Personnel Landmines, National Research Council, Washington, D.C., December 10, 1999.

Croll, M. 1998. The History of Landmines. Barnsley, U.K.: Leo Cooper.

DARPA (Defense Advanced Research Projects Agency). 2000a. Future Combat Systems Overview Abstract. DARPA Tactical Technology Office. Available on line: *http://www.darpa.mil/tto/programs/fcs.html.*

DARPA. 2000b. Microelectromechanical Systems Overview Abstract. DARPA Microsystems Technology Office. Available on line: *http://www.darpa.mil/mto/mems/overview.html.*

DARPA. 2000c. Affordable Moving Surface Target Engagement Information Overview Abstract. DARPA Special Projects Office. Available on line: *http://www.darpa.mil/spo/Programs/affordablemovingsurfacetarget.html.*

DARPA. 2000d. Sensor Information Technology Information Overview Abstract. DARPA Information Technology Office. Available on line: *http://www.darpa.mil/ito/personnel/skumar.html.*

DARPA. 2000e. Tactical Sensor Program Information Overview Abstract. DARPA Advanced Technology Office. Available on line: *http://www.darpa.mil/ato/programs/suo/baa98-18.html.*

Deagle, E.A. 2000. Military Alternatives to Landmines. Briefing by E. Deagle, Consultant, Vietnam Veterans of America Foundation, to the Committee on Alternative Technologies to Replace Anti-Personnel Landmines, National Research Council, Washington, D.C., May 24, 2000.

DOD (U.S. Department of Defense). 1999. Annual Report to the President and the Congress. Washington, D.C.: U.S. Department of Defense, Office of the Secretary of Defense. Available on line: *http://www.dtic.mil/execsec/adr1999.*

DOD. 2000. Progress on Landmine Alternatives: Report to Congress. April 1, 2000. Washington, D.C.: U.S. Department of Defense, Office of the Under Secretary of Defense (Acquisition, Technology, and Logistics).

DOS (U.S. Department of State). 2000. U.S. Humanitarian Demining Program Fact Sheet, April 1, 2000. Available on line: *http://www.state.gov/www/global/arms/pm/hdp/fs_000401_demine.html.*

Etienne Lacroix Defense. 1998. Sphinx: Perimeter Defence System. Paris, France: Etienne Lacroix Defense.

Fenton, G. 1999. Joint Non-Lethal Weapons Program. Briefing by G. Fenton, Director, Joint Non-Lethal Weapons Directorate, to a subcommittee of the Committee on Alternate Technologies to Replace Antipersonnel Landmines, Quantico, Virginia, December 21, 1999.

Gard, R.G. Jr. 1999. Alternatives to Antipersonnel Landmines. Vol. 1., no 1, in the VVAF Monograph Series. Washington, D.C.: Vietnam Veterans of America Foundation.

Garwin, R., and R. Sherman. 2000. Elimination of the Long-term Humanitarian Hazard of Stockpiled Antipersonnel Landmines. Briefing by R.

Garwin and R. Sherman, State Department Arms Control and Nonproliferation Advisory Board, to the Committee on Alternative Technologies to Replace Antipersonnel Landmines, Irvine, California, January 13, 2000.

Grayson, T. 2000. Affordable Moving Surface Target Engagement (AMSTE) Program. Briefing by T. Grayson, Defense Advanced Research Projects Agency (DARPA) Program Manager, AMSTE, to a subcommittee of the Committee on Alternative Technologies to Replace Antipersonnel Landmines, Arlington, Virginia, February 17, 2000.

Haseltine, E. 2000. Nonmateriel Alternatives. Briefing by E. Haseltine, Senior Vice President and Chief Scientist, Research and Development, Walt Disney Imagineering, to the Committee on Alternative Technologies to Replace Antipersonnel Landmines, Irvine, California, January 13, 2000.

Human Rights Watch. 2000. Landmine Monitor Executive Summary 2000: Toward a Mine-Free World. Washington D.C.: Human Rights Watch.

ICBL (International Campaign to Ban Landmines). 2000. A Brief History of the International Campaign to Ban Landmines. Available on line: *http://www:icbl.org*.

ICRC (International Committee of the Red Cross). 1949. Convention (iv) Relative to the Protection of Civilian Persons in the Time of War. Geneva, Switzerland: International Committee of the Red Cross. Available on line: *http://www.icrc.org/ihl.nsf*.

Irish, W. 2000. Nonlethal Bounding Munitions and Canister-Launched Area-Denial System. Briefing by W. Irish, Project Manager, Mines, Countermine, and Demolitions, to a subcommittee of the Committee on Alternative Technologies to Replace Antipersonnel Landmines, Picatinny, New Jersey, February 10, 2000.

Jane's. 1997. Legislation and the Landmine. Jane's Intelligence Review Special Report 16. Surrey, UK: Jane's Information Group.

JCS (Joint Chiefs of Staff). 1999. Joint Doctrine for Barriers, Tactics, and Mine Warfare. JCS Pub 3-15, February 24, 1999. Washington, D.C.: U.S. Department of Defense, Office of the Director, Joint Staff.

Kaminski, P.G. 1996. Presentation by P. Kaminski, Undersecretary of Defense for Acquisition and Technology, to American Defense Preparedness Association, Munitions Executive Summit in Fairfax, Virginia, September 17, 1996.

Kern, P. 1999. Army Futures. Briefing by P. Kern, Military Deputy, Office of the Assistant Secretary of the Army (Acquisition, Logistics, and Technology), to the Board on Army Science and Technology, Washington, D.C., December 2, 1999.

Kolasinski, E.M.1999. The Psychological Effects of Anti-Personnel Landmines: A Standard to Which Alternatives Can Be Compared. Engineering Psychology Laboratory Report 99-2. West Point, NY: U.S. Military Academy.

Krepinevich, A. 1994. Cavalry to Computer: The Pattern of Military Readiness. National Interest 37: 31–36.

Kumar, S. 2000. Sensor Information Technology. Briefing by S. Kumar, Program Manager, Sensor Information Technology Program, Defense Advanced Research Projects Agency (DARPA), to a subcommittee of the Committee on Alternatives Technologies to Replace Antipersonnel Landmines, Arlington, Virginia, March 13, 2000.

Matheson, M. 1999. Ottawa Convention and Convention on Conventional Weapons (Amended Mines Protocol). Briefing by M. Matheson, Principal Deputy Legal Advisor, U.S. Department of State, to the Committee on Alternatives Technologies to Replace Antipersonnel Landmines, National Research Council, Washington, D.C., December 9, 1999.

Metz, S. 2000. Armed Conflict in the 21st Century: The Information Revolution and Post-Modern Warfare. Carlisle, Pa.: U.S. Army War College, Strategic Studies Institute.

Morelli, W. 2000. APL-A Track III. Briefing by W. Morelli, Project Officer, Track III, to a subcommittee of the Committee on Alternative Technologies to Replace Antipersonnel Landmines, Arlington, Virginia, October 10, 2000.

New York Times. 1996. An Open Letter to President Clinton. The New York Times, April 3, 1996, p. A9.

News Report. 1963. The Centennial Convocation Address of President John F. Kennedy. National Academy of Sciences/National Research Council. News Report 13(6): 81–86.

NRC (National Research Council). 1997. Technology for the United States Navy and Marine Corps, 2000–2035. Volumes 2 and 5. Naval Studies Board. Washington, D.C.: National Academy Press.

O'Hanlon, M. 2000. Technological Change and the Future of Warfare. Washington, D.C.: Brookings Institution Press.

Patierno, P. 2000a. U.S. Humanitarian Demining Program. Remarks by P. Patierno, Director, Office of Humanitarian Demining Program, Bureau of Political Military Affairs, U.S. Department of State, April 2, 2000. Available on line: *http://www.state.gov/www/policy_remarks/2000/000402_patierno_hdp.html*.

Patierno, P. 2000b. Landmines: Human Rights and National Security. Address by P. Patierno, Director, Office of Humanitarian Demining Program, Bureau of Political Military Affairs, U.S. Department of State to Landmines: Human Rights and National Security, a forum sponsored by the National Committee on American Foreign Policy, New York, May 15, 2000. Available on line: *http://www.state.gov/www/policy_remarks/2000/000515_patierno_nyc.html*.

Persau, E. 2000. Nonlethal Mine Alternatives. Briefing by E. Persau, Office of the Project Manager, Mines, Countermine, and Demolitions, to a subcommittee of the Committee on Alternative Technologies to Replace Antipersonnel Landmines, Crystal City, Virginia, February 25, 2000.

Rieser, T. 1999. Congressional Perspective. Briefing by T. Reiser, Staff Member, Office of Senator Patrick Leahy (D-Vt), to the Committee on Alternative Technologies to Replace Antipersonnel Landmines, National Research Council, Washington, D.C., November 16, 1999.

Rigby, R. 1999. The U.S. Army and Future Warfare. Briefing by R. Rigby, Deputy Commanding General (Futures), U.S. Army Training and Doctrine Command, to the Committee on Alternative Technologies to Replace Antipersonnel Landmines, National Research Council, Washington, D.C., December 10, 1999.

Robbins, P. 1997. The Bomb Brothers. Civil War Times 36(4): 40–47.

Rosamilia, J. 2000. Nonself-Destruct Antipersonnel Landmine Alternative. Briefing by J. Rosamilia, Office of the Project Manager, Mines, Countermine, and Demolitions, Picatinny Arsenal, to the Committee on Alternative Technologies to Antipersonnel Landmines, Irvine, California, January 13, 2000.

Rossiter, C. 1999. Currently Available and Near-Term Alternatives to Land Mines to Achieve Combined Forces Command Objectives in Korea. Briefing by C. Rossiter, Vietnam Veterans of America Foundation, to the Committee on Alternative Technologies to Replace Antipersonnel Landmines, National Research Council, Washington, D.C., December 10, 1999.

Roy, R.L., and S. Friesen. 1999. Historical Uses of Antipersonnel Landmines: Impact on Land Force Operations. Research Note 9906. Kingston, Ontario: Operational Research Division, Canadian Department of National Defence.

Samuels, R.N. and N.D. Supola. 2000. Infrared cooled and uncooled staring sensor MTO. Army RD&A Jan / Feb, pp. 5–7.

Scales, R.H. 1999. Future Warfare Anthology. Carlisle, Pa.: U.S. Army War College.

Scales, R.H. 2000. Future Warfare. Discussion between MG Robert H. Scales, U.S. Army War College, and a subcommittee of the Committee on Alternative Technologies to Replace Antipersonnel Landmines, National Resesarch Council, Washington, D.C., March 28, 2000.

Schneck W.C. 1998. The Origins of Military Mines: Part I. Engineer (58): 50–52.

Shinseki, E. 2000. The Army transformation: a historic opportunity. Army, 50 (10): 21–30.

Strano, M.A. 2000. Raptor System Overview. Briefing by M. Strano, Office of the Project Manager, Mines, Countermine, and Demolitions, to a subcommittee of the Committee on Alternative Technologies to Replace Antipersonnel Landmines, Picatinny Arsenal, New Jersey, February 10, 2000.

Tilelli, J. H. 1999. Testimony before the House National Security Committee by J.H. Tillelli, Commander in Chief United Nations Command/Combined Forces Command and Commander United States Forces Korea. Congressional Record. 1999 106th Congress. 1st Sess. Vol. 145, No. 33. Available on line: *http://www.house.gov/hasc/testimony/106thcongress/99-03-03tilelli.htm.*

Tilelli, J.H. 2000. Requirements for Mines in Korea. Discussion with the Committee on Alternative Technologies to Replace Antipersonnel Landmines, National Research Council, Washington D.C., July 13, 2000.

Troxell, J.F. 1999. The Importance of Anti-Personnel Landmines in Korea. Briefing by J. Troxell, Director of National Security Studies, U.S. Army War College, to the Committee on Alternative Technologies to Replace Antipersonnel Landmines, National Research Council, Washington, D.C., December 10, 1999.

Toffler, A., and H. Toffler. 1993. War and Anti-War: Survival at the Dawn of the 21st Century. Boston, Mass.: Little Brown and Company.

U.S. Air Force. 1998. Global Engagement: A Vision for the 21st Century Air Force. Washington, D.C.: Headquarters Department of the Air Force. Available on line: *http://www.xp.hq.af.mil/xpx/21/nuvis.htm.*

U.S. Army. 1993. Operations. Field Manual 100-5. Washington, D.C.: Headquarters, Department of the Army.

U.S. Army. 1994. Combined Arms Obstacle Integration. Field Manual 90-7. Washington, D.C.: Headquarters, Department of the Army.

U.S. Army. 1997a. Army Vision 2010. Washington, D.C.: Headquarters, Department of the Army. Also available on line: *http://www.army.mil/2010/introduction.htm.*

U.S. Army. 1998a. 1997 Annual Review: Army Research Laboratory. Adelphi, Md.: Department of the Army, Army Research Laboratory.

U.S. Army. 1998b. Mine/Countermine Operations. Field Manual 20-32. Washington, D.C.: Headquarters, Department of the Army.

U.S. Army. 2000. Combined Arms Breaching Operations Field Manual 3-34.2. Washington, D.C.: Headquarters, Department of the Army.

U.S. Embassy Beijing. 1999. Analysis of Unrestricted Warfare. Part I, by People's Liberation Army (PLA) Colonels. Available on line: *http://www.usembassy-china.org.cn/english/sandt/unreswl.htm.*

U.S. Marine Corps. 1997. Operational Maneuver from the Sea: A Concept for the Projection of Naval Power Ashore. Washington, D.C.: Department of the Navy.

U.S. Navy. 1995. Forward...from the Sea. Washington, D.C.: Headquarters, Department of the Navy.

Washington Post. 2000. Postwar Review Found Far Fewer Serb Weapons Hit in Kosovo. Washington Post, May 9, 2000, p. 17.

White House. 1994. U.S. Policy on a Landmine Control Regime. Press release, September 26, 1994. Washington D.C.: Office of the Press Secretary.

White House. 1996. U.S. Announces Anti-Personnel Landmine Policy. Press release, May 16, 1996. Washington, D.C.: Office of the Press Secretary.

White House. 1997a. United States Announces Next Steps on Anti-Personnel Landmines. Press release, January 17, 1997. Washington, D.C.: Office of the Press Secretary.

White House. 1997b. U.S. Requirements for Landmines in Korea. Press release, September 17, 1997. Washington, D.C.: Office of the Press Secretary.

White House. 1999. A National Security Strategy for a New Century. December 1999. Washington, D.C.: White House.

Witkowsky, A. 1999. U.S. Antipersonnel Landmine Policy: Presidential Decision Directive. Briefing by A. Witkowsky, Director, Defense Policy and Arms Control, National Security Council Staff, to the Committee on Alternative Technologies to Replace Anti-Personnel Landmines, National Research Council, Washington, D.C., December 9, 1999.

Appendixes

Appendix A

Biographical Sketches of Committee Members

GEORGE BUGLIARELLO (chair) is chancellor of Polytechnic University, Brooklyn, New York. Dr. Bugliarello, former president (1973–1994) of Polytechnic, is an engineer and educator whose background ranges from biomedical engineering to fluid mechanics, computer languages, sociotechnology, and science policy. A member of the National Academy of Engineering (NAE) and the Council on Foreign Relations, and a founding fellow of the American Institute of Medical and Biological Engineering, he is a past president of Sigma Xi, the Scientific Research Society, and an honorary lifetime member of the National Association for Science, Technology and Society. He has served as both member and chair of several National Research Council committees and is currently a member of the Committee on Human Rights of the National Academy of Sciences, NAE, and the Institute of Medicine. Dr. Bugliarello's international experience includes consultant to UNESCO and U.S. member of the Science for Stability Steering Committee and the Science for Peace Steering Committee of the Scientific Affairs Division of NATO.

H. NORMAN ABRAMSON retired as the executive vice president of Southwest Research Institute in 1991 after 35 years of service in increasingly responsible positions. He has a B.S. in mechanical engineering and an M.S. in engineering mechanics from Stanford University and a Ph.D in engineering mechanics from the University of Texas-Austin. He is internationally known in the field of theoretical and applied mechanics, particularly for his expertise in the dynamics of contained liquids in astronautical, nuclear, and marine systems. Dr. Abramson is a past vice president and past governor of the American Society of Mechanical Engineers, a past director of the American Institute of Aeronautics and Astronautics, and an officer of the National Academy of Engineering (NAE). He was an elected member of the NAE Council from 1984 to1990. He has also served on a variety of NAE and National Research Council committees and panels, including the Committee on Science, Engineering, and Public Policy; U.S. National Committee on Theoretical and Applied Mechanics; Committee on Computational Mechanics; Committee on Earthquake Engineering Facilities; Ship Structure Committee; Research and Technology Coordinating Committee for FHWA; and the Committee on Technology Policy Options in a Global Economy. He has also been a member of advisory boards to government agencies and a consultant to several organizations.

THOMAS F. HAFER, senior program manager with Science and Technology Associates, has a long history of involvement in the search for alternatives to antipersonnel landmines by the U.S. Department of Defense and the U.S. Army. Mr. Hafer has an M.S. in electrical engineering. For 10 years, he managed programs at the Defense Advanced Research Projects Agency (DARPA) on the development of novel types of intelligent mines, communications for intelligent minefields, countermine and demining technologies, sensors, and situational awareness.

J. JEROME HOLTON, program manager and senior scientist with Defense Group Inc., earned his Ph.D. from Duke University in experimental physics. He has made a significant contribution to defense through a broad range of research and analyses, both technical and programmatic, related to the integration of spectral sensing technology and light detection and ranging (LIDAR). He has also worked on the integration of technologies for automatic target recognition algorithms to supplement other detection and identification systems on the battlefield, primarily for chemical and biological weapons.

LEE M. HUNT, former director of the National Research Council's (NRC) Naval Studies Board, has extensive experience with sea and landmine warfare, as well as explosive ordnance disposal, including sea and landmine countermeasures and explosive ordnance in World War II and the Korean Conflict. He has conducted more than 70 major studies in mine warfare, authored numerous papers on the subject, and worked to reduce the hazard of abandoned explosive ordnance since 1964. He is currently vice president for academic affairs of the Mine Warfare Association and a member of the Advisory Council for the College of Engineering, Florida State University.

RICHARD H. JOHNSON, a consultant who lives in Springfield, Virginia, has extensive experience with mine warfare. In his last assignment as an Army Colonel, capping 30 years of working on every aspect of ammunition, Mr. Johnson was the project manager for mines, countermine, and demolitions. During his career, he was responsible for managing the development and fielding of approximately 15 separate systems. Prior to his position as project manager, Mr. Johnson commanded the Fire Support Armaments Center providing full life-cycle engineering support for indirect-fire weapons, fire-control systems, mines and demolitions, and other selected systems, such as precision munitions. He recently was a member of a Defense Science Board panel that recommended improvements in unexploded ordnance remediation technology, and he works part time as a senior analyst on conventional arms control issues with Dyn-Meridian, Inc., Alexandria, Virginia.

K. SHARVAN KUMAR, professor of engineering, Brown University, has extensive experience and an international reputation in materials science, as well as technological expertise in a variety of applications in aerospace systems and defense-related technologies. Dr. Kumar has investigated the microstructure and mechanical properties of iron aluminides and their composites for potential use in passive mine countermeasures and led a team that evaluated the role of reinforcements in intermetallic and metallic matrices and identified innovative methods of processing these materials for specific applications. He has significant experience in the area of materials processing using combustion synthesis techniques and has participated in several Navy-funded projects on using exothermic reactions to produce a variety of metal-matrix composites. He has also worked on the development of materials using combustion synthesis for ballistic penetration resistance. Dr. Kumar holds several patents, one of which was licensed and used in the fabrication of the external fuel tank for the space shuttle.

RONALD F. LEHMAN II is director of the Center for Global Security Research at the U.S. Department of Energy's Lawrence Livermore National Laboratory. He is also chairman of the Governing Board of the International Science and Technology Center, an intergovernmental organization headquartered in Moscow, and a member of the U.S. Department of Defense Threat Reduction Advisory Committee. In 1995, he was appointed to the five-member President's Advisory Board on Arms Proliferation Policy. Ambassador Lehman was the director of the U.S. Arms Control and Disarmament Agency from 1989 to 1993. Earlier, he was Assistant Secretary of Defense for International Security Policy, chief U.S. negotiator on strategic offensive arms, and deputy assistant to the President for National Security Affairs. He has also served on the National Security Council staff as a senior director, on the professional staff of the U.S. Senate Armed Services Committee, and in Vietnam with the U.S. Army.

LARRY G. LEHOWICZ, Major General, U.S. Army (retired), and vice president of Quantum Research-International, is a graduate of the U.S. Army War College. A decorated combat leader, his current position in the civil sector is enhanced by his expertise in survivability analysis, developmental and operational testing and evaluation, and information operations. General Lehowicz's capstone position prior to the completion of his active military service was the command of the U.S. Army's Operational Test and Evaluation Command, an organization dedicated to ensuring that warfighting systems, information management systems, and other systems are operationally tested and evaluated. As the Deputy Chief of Staff for Combat Developments, U.S. Army Training and Doctrine Command, General Lehowicz was responsible for analyzing and recommending the most promising alternative conceptual and technological approaches for the future battlefield. He chaired the German, French, United Kingdom, and Israeli bilateral International Army Staff Talks, the mission of which was to ensure doctrinal compatibility and system interoperability on the future combined battlefield.

ALAN M. LOVELACE, a member of the National Academy of Engineering (NAE), is retired senior vice president of General Dynamics. Dr. Lovelace has had a long and distinguished career in both government and industry and has made significant contributions to U.S. preeminence in aerospace. His long experience with the National Academies includes a term as chair of the NAE. His areas of expertise are in aerospace materials, particularly the application of boron-reinforced and graphite-reinforced epoxies. Dr. Lovelace has a Ph.D. in chemistry, is a fellow and past president of the American Institute of Aeronautics and Astronautics, and is a member of the International Academy of Astronautics, Sigma Xi, and Phi Beta Kappa.

HARVEY M. SAPOLSKY, professor of public policy and organization, Political Science Department, Massachusetts Institute of Technology (MIT) and Director of the Security Studies Program at the MIT Center for International Studies, is an educator and advisor to governments. Dr. Sapolsky, who earned his Ph.D. in political economy and government, has worked in a number of public policy areas, notably defense, science, and health, examining the effects of technical change, institutional structure, and bureaucratic politics. He chaired the 1993 Policy Implications of Nonlethal Warfare

Technologies Conference at MIT and served as a member of both the Nonlethal Weapons Study Group for the Council on Foreign Relations and the Ethics and Health Policy Panel for the Hastings Center. He has served as a consultant to the Commission on Government Procurement, the Office of the Secretary of Defense, the Naval War College, the Office of Naval Research, the RAND Corporation, and the Applied Physics Laboratory at Johns Hopkins University. Dr. Sapolsky is a member of the Senior Review Group, Army Technology Seminar, the Assessment Panel for the U.S. Department of Defense Dual-Use Science and Technology Program and the Council on Foreign Relations. His most recent activity with the National Research Council was membership on the Committee on Risk Perception and Communication.

DANIEL R. SCHROEDER, Lieutenant General, U.S. Army (retired), has had a long and unique career in the military. He received his undergraduate degree in engineering from the U.S. Naval Academy and an M.S. in systems management from the Air Force Institute of Technology. He is a graduate of the U.S. Marine Corps Amphibious Warfare School and the U.S. Army War College. A decorated combat veteran, he counts among his many military assignments service as Deputy Assistant Chief of Engineers, Commandant of the U.S. Army Engineer School, and Deputy Commander in Chief of the U.S. Army Europe and Seventh Army. After completing his military service, General Schroeder was the senior vice president and general manager for modeling, simulation, and training for LORAL Federal Systems, now Lockheed Martin Information Systems. Since 1998, General Schroeder has been a consultant to several U.S. corporations.

MARION SCOTT is deputy director for sensors and national security at the Microsystems Science, Technology, and Components Center, Sandia National Laboratories. He attended Southern Methodist University where he received his Ph.D. in electrical engineering. He currently has programmatic responsibility for the development of sensors for a wide variety of applications, ranging from weapon state-of-health monitoring to the monitoring of industrial processes. Several of the sensors in this program area are being developed for nonproliferation applications. A major activity within his program area is the development of a miniature chemical analysis system for the detection of chemical warfare agents. Prior to assuming his current position, Dr. Scott was the manager of the Advanced Geophysical Technology Department and a staff member in the Optoelectronics Department at Sandia.

K. ANNE STREET is president and chief executive officer of Riverside Consulting Group, which specializes in the interaction of human factors and technology. In her previous position as president and chief operating officer of Geo-Centers, her projects included a ground-penetrating radar system capable of detecting landmines. Ms. Street has worked for several companies in her 27-year career, including DynCorp, Battelle Memorial Institute, The Parsons Corporation, and Fluor Engineers, Inc. She is a registered professional engineer and has an M.S. in ocean engineering and a B.S. in metallurgy and materials science. She is a member of the Board of Trustees of Aerospace Corporation and an active alumna of the Massachusetts Institute of Technology, including service as a member of the Visiting Committee for the Department of Materials Science and Engineering.

PATRICK H. WINSTON, Ford Professor of Artificial Intelligence and Computer Science at the Massachusetts Institute of Technology, earned his Ph.D. in computer science. He is involved in the study of how vision, language, and motor faculties account for intelligence and is working on applications of artificial intelligence enabled by learning, precedent-based reasoning, and common-sense problem solving. His new initiatives include the Human Intelligence Enterprise, which brings together and focuses research from several fields, including computer science, systems neuroscience, cognitive science, and linguistics. Dr. Winston also chaired the Naval Research Advisory Committee.

Appendix B

Committee Meetings

FIRST MEETING

November 15–16, 1999
Arlington, Virginia

Meeting objectives: introduce National Research Council procedures; complete administrative actions, including committee introductions and composition/balance/bias discussions for members, committee and report procedures, and committee administrative support methodology; discuss genesis of the study with congressional staff; discuss Statement of Task with sponsor; discuss draft report outline; discuss project plan and report realization; make writing assignments; receive overview briefing on strategic and tactical uses of antipersonnel landmines; determine objectives, location, and date of next committee meeting.

Presenters

Strategic & Tactical Landmine Usage Overview
Greg Bornhoft, project manager
BRTRC Technology Research Corporation

Sponsor Discussion Time
Peter O'Neill, sponsor representative
Office of the Secretary of Defense

Congressional Perspective
Tim Rieser, staff
Office of Senator Patrick Leahy

SECOND MEETING

December 9–11, 1999
Arlington, Virginia

Meeting objectives: complete composition/balance/bias discussions for committee members; achieve better understanding of compliance requirements for Ottawa Convention; achieve better understanding of both current and future strategic contexts for the use of antipersonnel landmines; consider previous efforts to provide alternative technologies to antipersonnel landmines; receive update on on-going efforts to provide alternative technologies to antipersonnel landmines; allow sufficient time for committee deliberations; discuss project plan and updated draft report outline; consider and approve Report Concept Draft; make writing assignments; determine objectives, location and date for the next committee meeting.

Presenters

The U.S. Army and Future Warfare
LTG Randall L. Rigby
U.S. Army Training and Doctrine Command

U.S. Antipersonnel Landmine Policy: Presidential Decision Directive
Anne Witkowsky
National Security Council Staff

Ottawa Convention and the Convention on Conventional Weapons (Amended Mines Protocol)
Michael J. Matheson
Principal Deputy Legal Advisor
U.S. Department of State

How and When the US Military Uses Landmines
LTC Michael W. Biering
Engineer Team Chief, ODCSOPS-FDD
Headquarters, Department of the Army

Joint Vision 2010
MG George Close
Director for Operational Plans and Interoperability (J-7)
Office of the Joint Chiefs of Staff

Antipersonnel Landmines Alternative Programs
Tom Altshuler, Program Manager
Defense Advanced Research Projects Agency

The Importance of Antipersonnel Landmines in Korea
COL John F. Troxell
Director of National Security Studies
U.S. Army War College

Currently Available and Near-Term Alternatives to Landmines to Achieve Combined Forces Command Objectives in Korea
Caleb Rossiter
Consultant, Vietnam Veterans of America Foundation

THIRD MEETING

January 13–15, 2000
Irvine, California

Meeting objectives: complete composition/balance/bias discussions for committee members, reaffirm understanding of Statement of Task; consider ongoing efforts to identify alternative technologies to antipersonnel landmines; consider nonmateriel alternatives; allow sufficient time for committee deliberations; discuss project plan and updated draft report outline; consider and approve report initial draft; finalize site visit objectives and locations; make writing assignments; determine objectives, location, and date of next committee meeting.

Presenters

LLNL Analysis of Antipersonnel Landmine/Alternative Utility
Robert J. Greenwalt, Jr., Engineer Analyst
Lawrence Livermore National Laboratory

Track 1–Technology Alternatives
John A. Rosamilia, Chief, Mine Division
Office of Project Manager, Mines, Countermine, and Demolitions

The Conservative Alternative to Persistent Antipersonnel Landmines
Richard L. Garwin
Chairman, State Department Arms Control and Nonproliferation Advisory Board
and
Robert Sherman
U.S. Department of State

U.S. Air Force Alternatives
Lt. Col. Charles Beason
Brooks Medical Center
and
Kirk E. Hackett
Air Force Research Laboratory

Non-Materiel Alternatives
Eric Haseltine, Senior Vice President and Chief Scientist
Walt Disney Imagineering

FOURTH MEETING

April 4–5, 2000
Washington, D.C.

Meeting objectives: reaffirm understanding of Statement of Task; discuss project plan and updated draft report outline; consider and mark up preliminary first full message draft; make writing assignments; determine objectives, location, and date of next committee meeting.

Presenter

Sponsor Discussion
Peter O'Neill, sponsor representative
Office of the Secretary of Defense

FIFTH MEETING

May 23–25, 2000
Washington, D.C.

Meeting Objectives: review report realization road map; discuss project plan and status; meet with CETS Editor; consider additional alternative ideas / systems; review / discuss preliminary concurrence draft; achieve initial consensus on conclusions and recommendations; make writing assignments, as necessary.

Presenters

U.S. Programs to Develop Alternatives to Antipersonnel Mines
Mark Hiznay, Senior Military Advisor
Human Rights Watch

Military Alternatives to Landmines
Edwin A. Deagle Jr., Consultant
Vietnam Veterans of American Foundation

Discussion on Editorial Procedures
Carol Arenberg, Editor
Commission on Engineering and Technical Systems
National Research Council

SIXTH MEETING

July 13–14, 2000
Washington, D.C.

Meeting Objectives: review report realization road map; discuss project plan and status; review / discuss concurrence draft; achieve consensus on conclusions and recommendations; make writing assignments, as necessary.

No presenters

SITE VISITS

Several site visits were made by subgroups of the committee to observe existing systems and research and development of new systems.

Quantico Marine Base

Quantico, Virginia
December 21, 1999

Participants

Joint Nonlethal Weapons Program
Col George Fenton, Director
Department of Defense, Joint Non-lethal Weapons Directorate

Joint Nonlethal Weapons Program
Ken Tiedge, Project Engineer
Department of Defense, Joint Non-lethal Weapons Directorate

Joint Non-Lethal Weapons Program
Kevin Stull, Project Engineer
Department of Defense, Joint Non-lethal Weapons Directorate

Vietnam Veterans of America Foundation

Washington, D.C.
February 1, 2000

Participant

Future Requirements for Antipersonnel Landmines
LTG Robert Gard, U.S. Army (ret.)
Consultant

Office of Project Manager for Mines, Countermine, and Demolitions

Picatinny Arsenal, New Jersey
February 10, 2000

Participants

Nonlethal Bounding Munitions and Canister Launched Area Denial System
COL Wilfred E. Irish
Project Manager for Mines, Countermines and Demolitions

Raptor System Overview
Marc A. Strano, Project Engineer
U.S. Army Tank-Automotive and Armaments
Command, Armament Research, Development and Engineering Center

Hornet / WAM
Richard Wagner
U.S. Army Armaments Research, Development and Engineering Center (ARDEC)

Logistics Management Institute

McLean, Virginia
February 17, 2000

Participant

Future Requirements for Antipersonnel Landmines
GEN William G. T. Tuttle, Jr., U.S. Army (ret.), President
Logistics Management Institute

Defense Advanced Research Projects Agency

Arlington, Virginia
February 17, 2000

Participant

Affordable Moving Surface Target Engagement (AMSTE) Program
Timothy P. Grayson
DARPA Special Projects Office

Office of Project Manager for Mines, Countermine, and Demolitions

Crystal City, Virginia
February 25, 2000

Participant

Nonlethal Mine Alternatives
Edward Persau
Project Officer

Defense Advanced Research Projects Agency

Arlington, Virginia
March 3, 2000

Participants

Sensor Information Technology
Sri Kumar
DARPA, Information Technology Office

Small Unit Operations: Tactical Sensors
Edward Carapezza
DARPA, Advanced Technology Office

Mobile Autonomous Robot Software
COL Mark Swinson
DARPA, Information Technology Office

Distributed Robotics Program
Ellison (Dick) C. Urban
DARPA, Microsystems Technology Office

Vietnam Veterans of America Foundation

Washington, D.C.
March 8, 2000

Participant

Future Requirements for Antipersonnel Landmines
LTG Robert Gard, U.S. Army (ret.)
Consultant

U.S. Army War College

Washington, D.C.
March 28, 2000

Participant

Future Requirements for Antipersonnel Landmines
MG Robert H. Scales, Jr.
Commandant

U.S. Army Night Vision and Electronic Sensors Directorate

Fort Belvoir, Virginia
March 31, 2000

Participants

NVESD Simulation Support
Mid Self, Chief, Virtual Experiments
Modeling and Simulation Division
U.S. Army Night Vision and Electronic Sensors Directorate

Potential Applications to Alternate Mine Solutions
William D'Amico
Advanced Munitions Concepts Branch
Weapons and Materials Research
Directorate, U.S. Army Research Laboratory

Uncooled Micro-camera Technology
Don Reago, Director, Science and Technology Division
U.S. Army Night Vision and Electronic Sensors Directorate

Launched Grapnel Hook
Larry Nee, Chief Countermine Division
Office of Project Manager for Mines, Countermine, and Demolitions

Sensor Applications for Antipersonnel and Landmine Alternatives and Beyond
Michael Jennings, Deputy
U.S. Army Night Vision and Electronic Sensors Directorate

Warrior Extended Battlespace Sensors
John Eicke
U.S. Army Research Laboratory

U.S. Department of Defense

Washington, D.C.
May 5, 2000

Participant

Future Requirements for Antipersonnel Landmines
Hans Mark
Director of Defense Research and Engineering

Office of Project Manager for Mines, Countermine, and Demolitions

Pentagon, Virginia
October 10, 2000

Participant

Track III Alternatives
William Morelli
Project Officer

Washington Team Meetings

A small number of committee members in the national capital region established a subgroup that met many times to address the evaluation methodology. Members of this group were Thomas F. Hafer, J. Jerome Holton, Lee M. Hunt, Richard H. Johnson, Larry Lehowicz, and K. Anne Street.

Appendix C

Current Types of U.S. Landmines

This appendix provides descriptions of currently available antipersonnel landmines (APL) and some antitank (AT) mines that are part of mixed systems (see Table C-1). If current presidential policy continues, all of the pure APL described here, with the exception of the Claymore, will be unavailable to the U.S. military after 2003, except in Korea. If Congress ratifies the Ottawa Convention, all APL, including mixed systems will become permanently unavailable to the U.S. military everywhere. The loss of the Area-Denial Artillery Munition (ADAM), the only APL that can be delivered by artillery, would create a significant gap in U.S. military capabilities.[1]

Several disadvantages of current mines might be eliminated with alternatives. These include an inability to distinguish between combatants and noncombatants; the need for launch platforms for remotely delivered mines; and fratricide among friendly forces.

FOR USE AGAINST DISMOUNTED THREATS

Dismounted soldiers can be targeted using either hand-emplaced APL or the family of scatterable mines (FASCAM). Hand-emplaced APL, which are labor intensive and time consuming to use, can be exploded by direct pressure, a trip wire, or command. They may be either buried in patterned minefields or laid on the surface randomly or in a pattern. These mines were designed for use in protective and tactical minefields either by themselves or in conjunction with hand-emplaced AT mines.

Scatterable APL, which are remotely delivered, are designed for use in protective and tactical minefields. These mines are scattered on the surface in a random distribution and are programmed to self-destruct or self-deactivate after a certain period of time following deployment. Each mine has a preengraved fragmentation warhead and is detonated by pulling on a thin trip wire. Because scatterable APL can be delivered remotely, they can be used deep in the battlespace.

The limitations of the delivery system have a direct bearing on the size and ease of emplacement of a minefield. For instance, it takes considerable artillery assets to deliver a large pure AT minefield because each 155-millimeter (mm) projectile contains only nine mines. High-performance and rotary-wing aircraft can deliver a relatively large number of mines but may be vulnerable to enemy fire.

Figure C-1 shows how various landmine systems would be used across a battlefield. Each system has a particular mission on the battlefield.

Hand-Emplaced Mines

The M14 (Figure C-2) is a nonself-destructing blast mine with a cylindrical plastic body (56 mm in diameter and 40 mm high) and a minimum metal content of approximately 2.36 grams in the firing pin. The M14 is fired when pressure (11.5 to 13.5 kilograms [kg]) is applied to the top surface. The main charge is 28.35 grams of tetryl. The design was chosen because of logistical considerations; because of their small size, a soldier can carry several M14 mines. U.S. forces use the M14 only in Korea, where metal washers have been added to meet detectability requirements of the Convention on Conventional Weapons (CCW).

The M16 (Figure C-3) is a nonself-destructing, cylindrical, metal-body mine (103 mm in diameter and 199 mm high, including the fuze) with a bounding fragmentation warhead. It weighs 3.5 kg and contains 450 grams of TNT. The M16 has a casualty radius of 27 to 30 meters, and the mine is detonated by a direct force of 3.6 to 9 kg or a pull of 1.4 to 4.5 kg on a trip wire.

Figure C-4 shows a pursuit denial munition, a self-destructing/self-deactivating APL. It is an adaptation of a single ADAM mine. The pursuit denial munition is a

[1] To fill this anticipated gap, the U.S. Department of Defense has proposed using the Remote Area-Denial Munition (RADAM) described in Chapter 6.

TABLE C-1 Current U.S. Mines

System Name	APL/ AT/ Mixed Non-mine	Self-destructing/ Self-deactivating	Lethal/ Nonlethal	Ottawa Compliant[a]	Dismounted Enemy		Mounted Enemy	
					Remotely Delivered	Hand Emplaced	Remotely Delivered	Hand Emplaced
M14	APL	N	L	N		X		
M16	APL	N	L	N		X		
Claymore (M18)	APL	N	L	Y		X		
Pursuit-Denial Munition (PDM)	APL	Y	L	N		X		
Ground-Emplaced Mine Scattering System (GEMSS)	Mix	Y	L	N	X		X	
Modular-Pack Mine System (MOPMS)	Mix	Y	L	N		X		X
Area-Denial Artillery Munition (ADAM)	APL	Y	L	N	X			
Remote Antiarmor Mine System (RAAMS)	AT	Y	L	Y			X	
Gator	Mix	Y	L	N	X		X	
Hornet / Wide Area Munition (WAM)	AT	Y	L	Y				X
Volcano (M87)	Mix	Y	L	N	X		X	
Volcano (M87A1)	AT	Y	L	Y			X	

Note: Systems in italics will not be available outside Korea after 2003 and will not be available anywhere once the Ottawa Convention is ratified.

FIGURE C-1 Landmine systems on the battlefield.

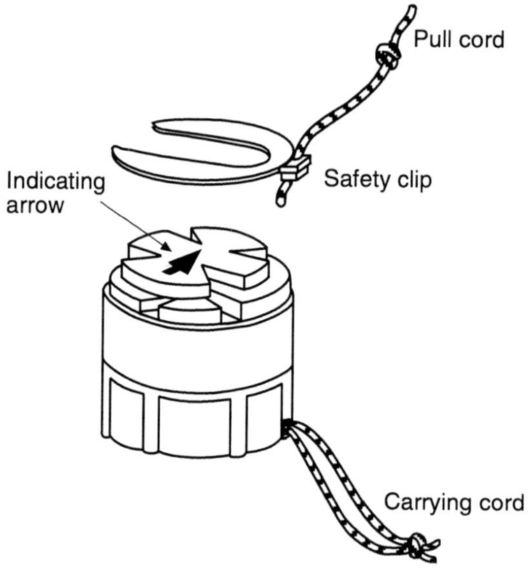

FIGURE C-2 M14.

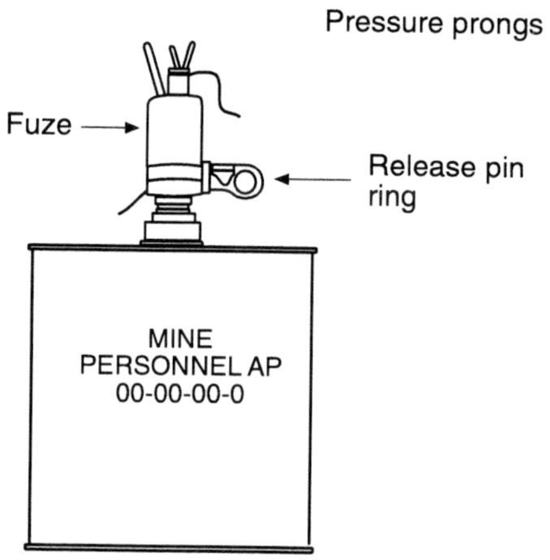

FIGURE C-3 M16.

hand-placed mine used by Special Operating Forces to inhibit enemy pursuit during a withdrawal from an operational area. Conceptually, this munition can be carried and used similarly to a hand grenade, but it has the capabilities of an APL. It has a factory-set self-destruct time of four hours. After placement, it operates exactly like an ADAM mine.

The M18 series, or Claymore mine (Figure C-5), is a non-self-destructing directional fragmentation mine detonated by 682 grams of composition C4. The Claymore projects 700 steel balls in a fan-shaped pattern in a 60-degree arc to a maximum height of 2 meters. The M18 can be activated in the command-detonation mode by an electric blasting cap inserted into the detonator well. The mine body is nonmetallic, and the steel balls are cast in the composition that comprises the front part of the mine. The lethal radius extends to 100 meters forward and 16 meters to the rear. The Claymore mine is issued with an electric initiation system (hand generator, wire, and electric blasting cap) and is now doctrinally detonated by command, although so-called "mechanical ambushes," using trip-wires, were common during the Vietnam War. Amended Protocol II to the CCW imposes restrictions on the use of trip wires to detonate Claymores; the

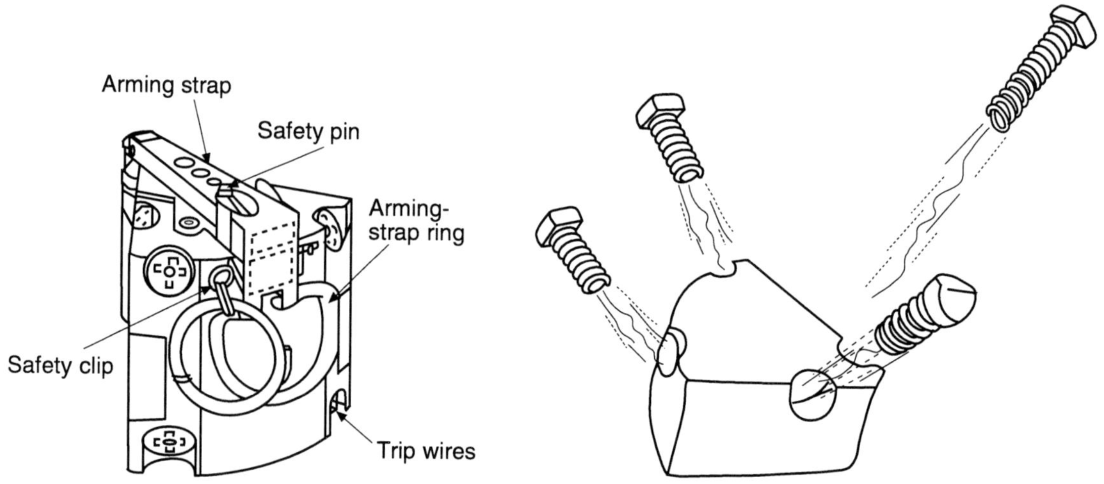

FIGURE C-4 Pursuit denial munition.

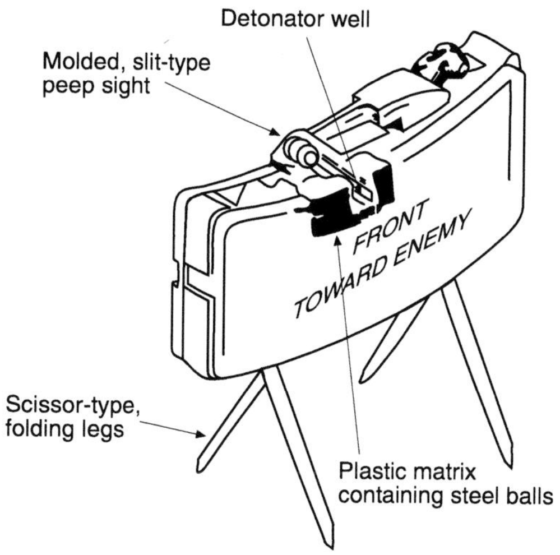

FIGURE C-5 M18 Claymore.

Ottawa Convention prohibits the use of trip wires with all mines. The command-detonated Claymore does not use a trip wire and is, therefore, permitted by both the CCW and the Ottawa Convention.

Remotely Delivered (Scatterable) Mines

The ADAM system (Figure C-6), a self-destructing/self-deactivating APL, consists of a 155-mm howitzer projectile that contains 36 APL. When the projectile time fuze fires at the scheduled time, the mines are ejected from the base of the projectile above the target area. Each mine is a 60-degree wedge with a radius of about 63 mm and a height of about 70 mm; a full cylinder inside the projectile holds six mines. Once the mines land on the ground, depending on their orientation, three or four 6-meter-long trip wires are deployed for each mine. The trip wires are thin, olive drab filaments that are difficult to see. Each mine has a spherical prefragmented kill mechanism with a charge of 21 grams of composition A5 explosive. When a force of .9 kg to 1.4 kg is exerted on the trip wire, the spherical warhead is projected upwards and explodes 0.6 to 2.4 meters above the ground; the lethal radius is 6 to 10 meters. ADAM mines have factory-set self-destruct times of either 4 or 48 hours; all of the mines in a projectile have the same self-destruct time.

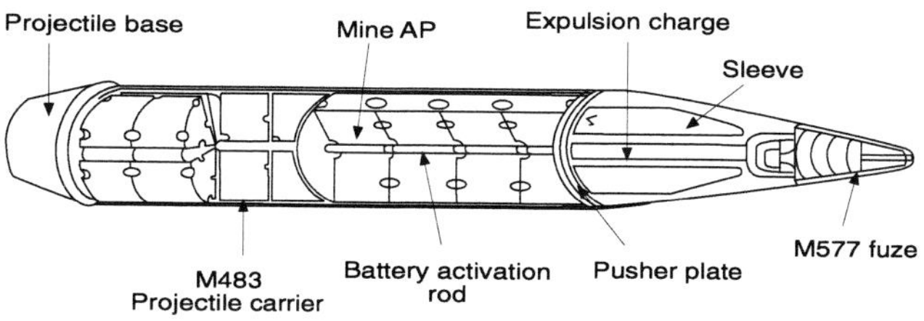

FIGURE C-6 ADAM mine projectile.

ADAM mines are used in conjunction with AT mines to delay dismounted breaches. In counterbattery fire, they are most effective against towed artillery. Because they are artillery-dispensed submunitions, it is difficult to use them in protective minefields close to friendly troops. The maximum range of the projectile is approximately 20 kilometers, and a significant number of artillery assets are required to create or cover a large minefield.

FOR USE AGAINST MOUNTED THREATS

The mines intended to destroy tanks or tracked vehicles are large, sometimes relatively easy to see munitions that can be countered in a variety of ways. Sometimes APL are used with AT mines to discourage an enemy from neutralizing the larger mines and breaching the minefield. Opinions differ about the efficacy of using APL to protect AT mines. With the exception of the Hornet/Wide Area Munition (WAM), which must be hand emplaced, all of these mines are remotely delivered. All AT mines are powered by reserve cells, have an electronic fuze or lens, contain a safe-and-arming device to provide environmental sensing, and are exploded by an electric detonator. AT mines are scattered on the surface and set to self-destruct at a designated time. They use magnetic-influence fuzes and the Miznay-Shardin effect (or self-forging fragment) to provide full-width coverage under armored targets. Because each artillery projectile contains only nine mines, a considerable amount of artillery is required to deliver a large AT minefield. High-performance or rotary-wing aircraft can deliver a relatively large number of mines but may be vulnerable to enemy fire.

Mixed systems combine APL and AT mines in a single delivery mode, carried by soldiers, ground vehicles, or aircraft. All of the mines have the same physical properties (right-circular cylinders, approximately 120 mm in diameter, 80 mm high, and weighing approximately 1.8 kg). The type of safe-and-arming device and is determined by the launch environment.

Hand-Emplaced Mines

The Hornet/WAM (Figure C-7) is an autonomous AT mine currently entering inventory in the hand-emplaced version. It weighs about 16 kg, is about 36 cm high and about 23 cm in diameter. After it is emplaced and armed, its seismic sensors can detect movement, and alert the mine to turn on its acoustic sensors to detect and classify a target. If an armored target approaches within 100 meters, a small

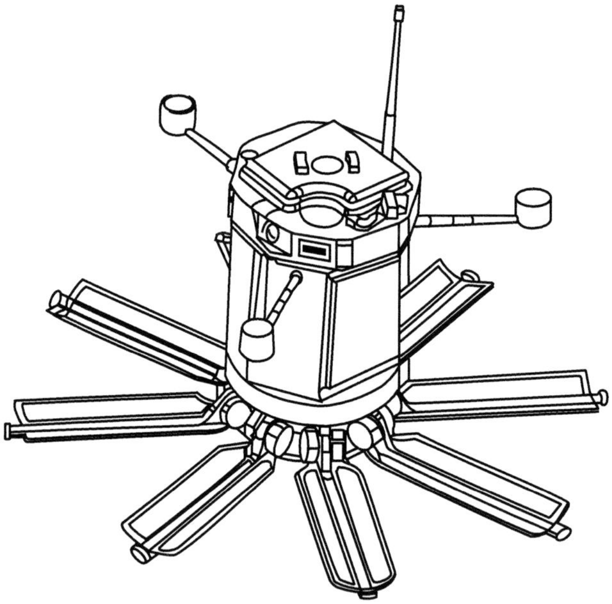

FIGURE C-7 Hornet/WAM.

submunition or sublet with an infrared sensor is launched over the target and fires an explosively formed projectile down into the engine compartment. The Hornet/WAMs has an antihandling feature that causes the mine to detonate when disturbed. The Hornet/WAMS is designed to operate for 30 days after emplacement and to then self-destruct. The sublet is similar to the sublets in the Air Force Sensor-Fuzed Weapon. Original plans called for the Hornet/WAM to be dispensed from a Volcano launcher or a deep-attack asset, such as the Multiple-Launch Rocket System (MLRS) or the Tactical Missile System (TACMS), in addition to hand emplacement. However, subsequent evolutions of requirements and the exigencies of the development program have precluded these options. Therefore, for the foreseeable future, Hornet/WAM will be hand emplaced, which sharply limits its utility.

Remotely Delivered (Scatterable) Mines

The Remote Antiarmor Mine System (RAAMS) (Figure C-8) consists of a 155-mm howitzer projectile containing nine AT mines. RAAMS is usually used with the ADAM, but can be used alone. Each mine in the projectile is a right-circular cylinder, 11.25 cm in diameter and 7.5 cm high, weighing 1.8 kg and containing about 0.78 kg of pressed explosive in the main charge. RAAMS have factory-set self-destruct times of either 4 or 48 hours. About 20 percent have an antihandling feature that causes them to explode when tilted at an angle of 17 degrees or more. RAAMS are designed to destroy mounted targets by perforating the underside of the vehicle. The mines have magnetic-influence fuzes designed to stop mounted vehicles when they are detonated, as intended, between the tracks; if they are detonated directly under a track, they may not stop the vehicle.

The Ground-Emplaced Mine-Scattering System (GEMSS/Flipper) was the first scatterable mine system fielded by the Army (RAAMS/ADAM was the second). Although GEMSS (Figure C-9) is now obsolete, the mines are still in the U.S. inventory. The GEMSS dispenser is a large trailer that can be towed by a variety of wheeled and tracked vehicles. A relatively large tactical minefield can be laid in areas controlled by friendly forces. The dispenser holds 800 mines and can lay mines in 26, 34, or 60-meter wide strips in five different densities. The mines come in sleeves of five mines each. The operator sets self-destruct times at either 5 or 15 days. Because the system is unwieldy to tow and operators were vulnerable to small arms fire, the utility of the system was limited. Flipper is a relatively small,

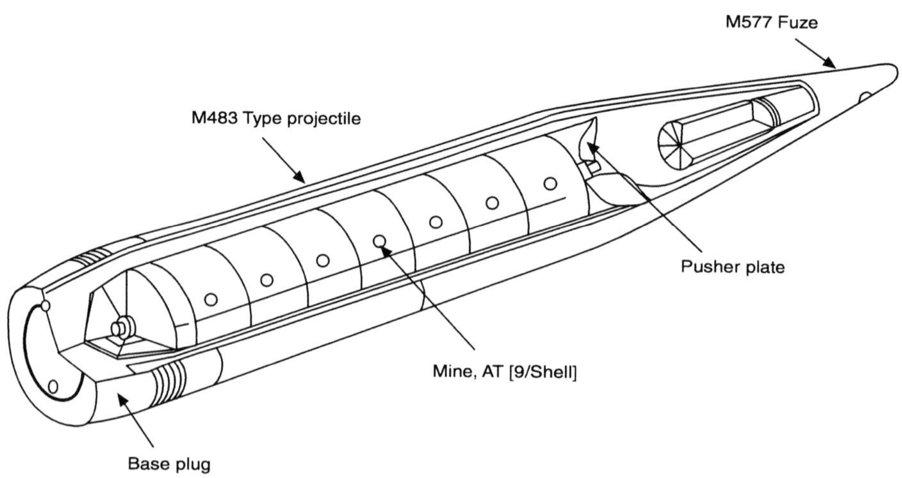

FIGURE C-8 RAAMS projectile.

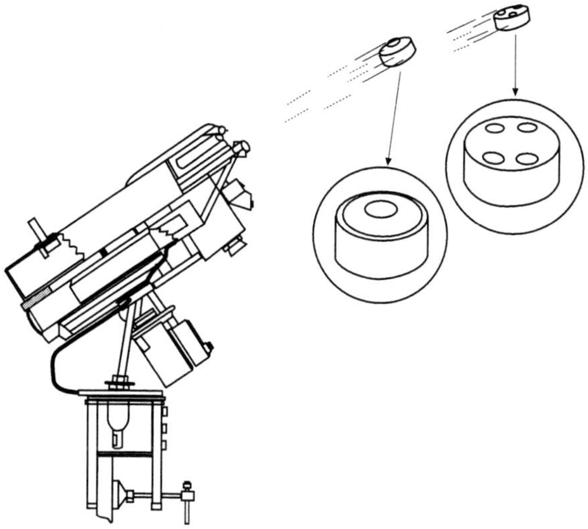

FIGURE C-9 GEMSS system.

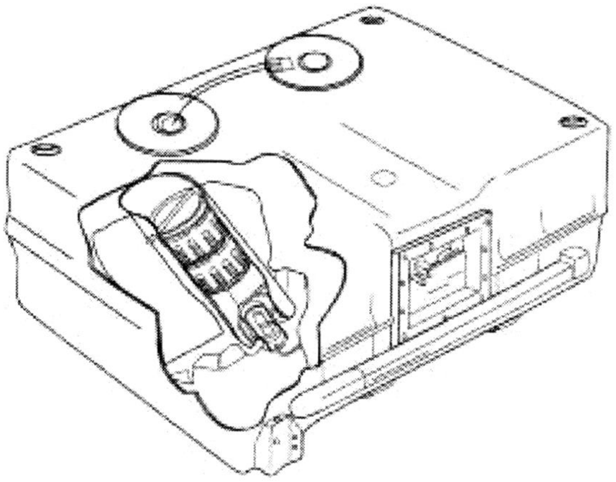

FIGURE C-10 MOPMS.

manual, auxiliary dispenser system designed to dispense GEMSS mines. Flipper can be mounted on the rear of a vehicle and was used mostly to place small point or protective minefields close to friendly troops.

The Modular-Pack Mine System (MOPMS) (Figure C-10) is normally used to fill in gaps in larger minefields and to establish point minefields. More systems are required to construct a large tactical minefield than are normally available to a single unit. Each man-portable dispenser contains 17 AT mines and 4 APL. The mines are dispensed on command by a signal sent either by wire or from a special radio transmitter. If the mines are not dispensed, the dispenser can be relocated and used elsewhere. The mines have a set self-destruct time of 4 hours. Of all currently available scatterable mines, MOPMS is the only system capable of extending the self-destruct time (up to three times). In addition, MOPMS can be detonated by command if the tactical situation dictates.

Gator mines (Figure C-11) are delivered by high-performance aircraft and can be used anywhere on the battlefield a tactical aircraft can reach. The Air Force and Navy versions of Gator differ only in the number of mines contained in the tactical munitions dispensers used to deliver them. The Air Force dispenser, the CBU-89/B, contains 72 AT mines and 22 APL; the Navy dispenser, the CBU-78/B, contains 45 AT mines and 15 APL. All of the mines have square-shaped, aeroballistic protective casings designed to aid dispersion after the dispenser opens above the target area. Prior to release, the pilot sets self-destruct times of 4 hours, 48 hours, or 15 days. Although Gator can be used in close combat situations, it is normally used against deep or interdiction targets.

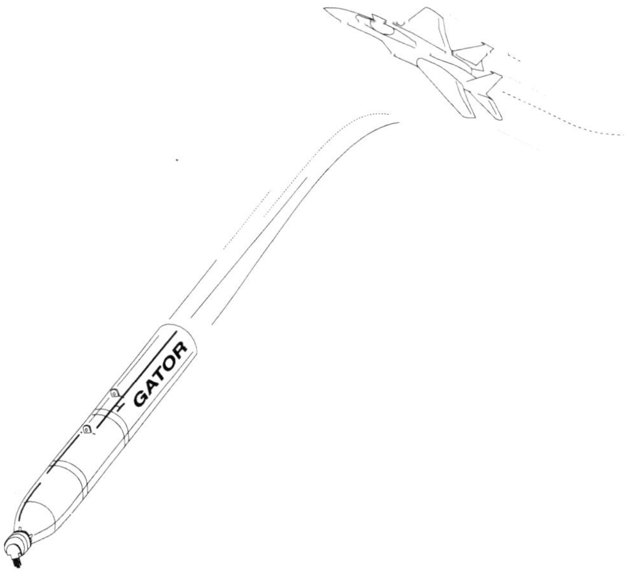

FIGURE C-11 Gator projectile.

FIGURE C-12 Volcano delivery system.

The Volcano system (M87) (Figure C-12) replaced the older M56 helicopter-delivered AT mine system and the GEMSS/Flipper system. The M87 is a mixed system and the M87A1 is a pure AT system. Volcano dispensers can be mounted on several tracked or wheeled vehicles or on the UH-60 Blackhawk helicopter. The system is made up of the launcher rack and dispenser-control unit, vehicle-specific mounting hardware, and mine canisters, each of which holds six mines. A completely loaded dispenser holds 160 canisters, or 960 mines. The mines are placed in a uniform density of approximately one mine per linear meter over a linear distance of one kilometer. Self-destruct times of 4 hours, 48 hours, or 15 days are set at launch time. The Volcano system can be used to emplace protective and tactical minefields anywhere on the battlefield reachable by the dispensing vehicles.

Appendix D

Value of Antipersonnel Landmines in Unprotected Mixed Minefields

As described in Chapter 3, one of the uses for APL is to protect AT mines from dismounted enemy soldiers who could manually disable or move the unprotected antitank mines (i.e., breach the minefield) and allow vehicles to pass through the minefield. *Breaching operations* are defined as operations that allow maneuvers to continue despite the presence of obstacles. *Obstacle breaching* is the use of a combination of tactics and techniques to advance an attacking force to the far side of an obstacle that is normally covered by fire. Obstacle breaching is perhaps the single, most difficult combat task a force must perform.

Breaching is a synchronized combined-arms operation under the control of a maneuver commander. Breaching operations begin when friendly forces detect an obstacle and begin to apply breaching fundamentals, and they end when battle handover has occurred between the follow-on forces and the unit conducting the breaching operation (U.S. Army, 2000). See Box D-1 for a more complete explanation of breaching.

Modern, remotely delivered, U.S. AT mines remain on the ground surface, in most cases, clearly visible, to an approaching enemy. By doctrine, direct-fire and indirect-fire weapons are the most effective deterrents to enemy breaches, but tactical minefields are often beyond the range of observers and protective fire, especially in economy-of-force sectors. Under these circumstances, APL are perceived to be particularly valuable in slowing the advance of dismounted enemy forces.

This perception was called into question in the course of this study, and the committee concluded that the data used to validate the effectiveness of APL in breaching operations should be updated. Only one document available to the committee, the final report from the Modeling and Analysis Group, APL Alternatives Study, found in the *Examination of the Battlefield Utility of Antipersonnel Landmines and the Comparative Value of Proposed Alternatives*, contained analytical data (Greenwalt and Magnoli, 1997). That report concluded that the use of APL with trip wires significantly increases the time required to breach a minefield. Up to 80 minutes was attributed to the trip wires.

However, the committee noted that the report scenario was based only on the use of hand-thrown grapnel hooks[1] to clear trip wires. The study did not describe or analyze other possible methods of clearing trip wires, such as the launched grapnel hook and man-portable projected line charges (e.g., the Small Projected Line Charge [SAPLIC]) or the Antipersonnel Obstacle Breaching System [APOBS]).

The U.S. Army's launched grapnel hook, based on technology available on the commercial market, was tested, type classified, and fielded in 1995. According to the manufacturer it takes about 30 seconds to put the launched grapnel hook into operation and fire it (telephone conversation between Mr. Steve Adleman, SAA International, LTD. and Richard Johnson, member of the Committee on Alternative Technologies to Replace Antipersonnel Landmines, April 24, 2000). It may take up to two minutes to retrieve the device, depending on terrain and foliage, and it is desirable to make at least two passes to ensure that all of the trip wires are cleared. Thus, it would appear that using a launched grapnel, a soldier could clear trip wires in about five minutes.

According to *Jane's Mines and Mine Clearance, Fourth Edition, 1999*, man-portable projected line charges are manufactured in many countries and are available on the world market (Jane's, 1999). A projected line charge can cut trip wires, clear a narrow path immediately under the deployed line charge, and leave a trace on the ground to use as a guide while the breach lane is cleared. According to the manufacturer[2] of the SAPLIC, a one-man-portable projected line

[1] Use of the grapnel is the only unique task in breaching an AT minefield containing APL with trip wires; all other steps in breaching an AT minefield are identical regardless of its composition.

[2] The Ensign-Bickford Company, Simsbury, Connecticut, manufactures SAPLIC and a larger man-portable line charge, APOBS. Both systems are capable of cutting trip wires and clearing breach lanes during tactical assaults.

BOX D-1
Fundamentals of U.S. Breaching Operations

Suppress, obscure, secure, reduce, and assault are the breaching fundamentals that must be applied to ensure success when breaching against a defending enemy. These fundamentals will always apply, but they may vary based on the specific battle-space situation (mission, enemy, terrain, troops, time available, and civilian considerations).

Suppress - Suppression is a tactical task used to employ direct or indirect fires or an electronic attack on enemy personnel, weapons, or equipment to prevent or degrade enemy fires and observation of friendly forces. The purpose of suppression during breaching operations is to protect forces reducing and maneuvering through an obstacle.

Obscure - Obscuration must be employed to protect forces conducting obstacle reduction and the passage of assault forces. Obscuration hampers enemy observation and target acquisition and conceals friendly activities and movement. Obscuration smoke deployed on or near the enemy's position minimizes its vision. Screening smoke employed between the reduction area and the enemy conceals movement and reduction activities. It also degrades enemy ground and aerial observations. Obscuration must be carefully planned to provide maximum degradation of enemy observation and fires, but it must not significantly degrade friendly fires and control.

Secure - Friendly forces secure the reduction area to prevent the enemy from interfering with obstacle reduction and the passage of the assault force through the lanes created during the reduction. Security must be effective against outposts and fighting positions near the obstacle and against overwatching units, as necessary. The far side of the obstacle must be secured by fires or be occupied before attempting any effort to reduce the obstacle. The attacking unit's higher headquarters has the responsibility to isolate the breach area by fixing adjacent units, attacking enemy reserves in depth, and providing counter fire support.

Reduce - Reduction is the creation of lanes through or over an obstacle to allow an attacking force to pass. The number and width of lanes created varies with the enemy situation, the assault force's size and composition, and the scheme of maneuver. The lanes must allow the assault force to rapidly pass through the obstacle. The breach force will reduce, proof (if required), mark, and report lane locations and the lane-marking method to higher headquarters. Follow-on units will further reduce or clear the obstacle when required. Reduction cannot be accomplished until effective suppression and obscuration are in place, the obstacle has been identified, and the point of breach is secure.

Assault - A breaching operation is not complete until friendly forces have assaulted to destroy the enemy on the far side of the obstacle that is capable of placing or observing direct and indirect fires on the reduction area and battle handover with follow-on forces has occurred (if desired).

Source: U.S. Army, 2000.

charge capable of clearing a path up to 80 meters long, the setup and employment time is no more than 60 seconds. Other devices listed in *Jane's* should have similar employment times.

Considering the confusion and other conditions on the battlefield, the so-called "fog of war," it would probably take longer to use a launched grapnel or a man-portable line charge in a real engagement than in an operational test. To compensate for the inherent confusion of combat, the committee notionally doubled the times claimed by the manufacturer and concluded that APL would provide, at most, approximately 10 minutes of additional protection for AT mines, if the enemy forces were properly equipped. In fact, the greatest time advantage provided by APL would be during a dismounted breach conducted under fire. In this situation, however, protective fires would cause a much greater delay than the APL.

REFERENCES

Greenwalt, B., and D. Magnoli. 1997. Examination of the Battlefield Utility of Antipersonnel Landmines and the Comparative Value of Proposed Alternatives. Livermore, Calif.: Lawrence Livermore National Laboratory.

Jane's. 1999. Mines and Mine Clearance, 4th ed., Surrey, U.K.: Jane's Information Group.

U.S. Army. 2000. Combined Arms Breaching Operations Field Manual 3-34.2. Washington, D.C.: Headquarters, Department of the Army.

Appendix E

The Ottawa Convention and Amended Protocol II of the Convention on Conventional Weapons

CONVENTION ON THE PROHIBITION OF THE USE, STOCKPILING, PRODUCTION AND TRANSFER OF ANTI-PERSONNEL MINES AND ON THEIR DESTRUCTION (OTTAWA CONVENTION)

Preamble

The States Parties,

Determined to put an end to the suffering and casualties caused by anti-personnel mines, that kill or maim hundreds of people every week, mostly innocent and defenceless civilians and especially children, obstruct economic development and reconstruction, inhibit the repatriation of refugees and internally displaced persons, and have other severe consequences for years after emplacement,

Believing it necessary to do their utmost to contribute in an efficient and coordinated manner to face the challenge of removing anti-personnel mines placed throughout the world, and to assure their destruction,

Wishing to do their utmost in providing assistance for the care and rehabilitation, including the social and economic reintegration of mine victims,

Recognizing that a total ban of anti-personnel mines would also be an important confidence-building measure,

Welcoming the adoption of the Protocol on Prohibitions or Restrictions on the Use of Mines, Booby-Traps and Other Devices, as amended on 3 May 1996, annexed to the Convention on Prohibitions or Restrictions on the Use of Certain Conventional Weapons Which May Be Deemed to Be Excessively Injurious or to Have Indiscriminate Effects, and calling for the early ratification of this Protocol by all States which have not yet done so,

Welcoming also United Nations General Assembly Resolution 51/45 S of 10 December 1996 urging all States to pursue vigorously an effective, legally-binding international agreement to ban the use, stockpiling, production and transfer of anti-personnel landmines,

Welcoming furthermore the measures taken over the past years, both unilaterally and multilaterally, aiming at prohibiting, restricting or suspending the use, stockpiling, production and transfer of anti-personnel mines,

Stressing the role of public conscience in furthering the principles of humanity as evidenced by the call for a total ban of anti-personnel mines and recognizing the efforts to that end undertaken by the International Red Cross and Red Crescent Movement, the International Campaign to Ban Landmines and numerous other non-governmental organizations around the world,

Recalling the Ottawa Declaration of 5 October 1996 and the Brussels Declaration of 27 June 1997 urging the international community to negotiate an international and legally binding agreement prohibiting the use, stockpiling, production and transfer of anti-personnel mines,

Emphasizing the desirability of attracting the adherence of all States to this Convention, and determined to work strenuously towards the promotion of its universalization in all relevant fora including, inter alia, the United Nations, the Conference on Disarmament, regional organizations, and groupings, and review conferences of the Convention on Prohibitions or Restrictions on the Use of Certain Conventional Weapons Which May Be Deemed to Be Excessively Injurious or to Have Indiscriminate Effects,

Basing themselves on the principle of international humanitarian law that the right of the parties to an armed conflict to choose methods or means of warfare is not unlimited, on the principle that prohibits the employment in armed conflicts of weapons, projectiles and materials and methods of warfare of a nature to cause superfluous injury or unnecessary suffering and on the principle that a distinction must be made between civilians and combatants,

Have agreed as follows:

Article 1
General obligations

1. Each State Party undertakes never under any circumstances:

a) To use anti-personnel mines;

b) To develop, produce, otherwise acquire, stockpile, retain or transfer to anyone, directly or indirectly, anti-personnel mines;

c) To assist, encourage or induce, in any way, anyone to engage in any activity prohibited to a State Party under this Convention.

2. Each State Party undertakes to destroy or ensure the destruction of all anti-personnel mines in accordance with the provisions of this Convention.

Article 2
Definitions

1. "Anti-personnel mine" means a mine designed to be exploded by the presence, proximity or contact of a person and that will incapacitate, injure or kill one or more persons. Mines designed to be detonated by the presence, proximity or contact of a vehicle as opposed to a person, that are equipped with anti-handling devices, are not considered anti-personnel mines as a result of being so equipped.

2. "Mine" means a munition designed to be placed under, on or near the ground or other surface area and to be exploded by the presence, proximity or contact of a person or a vehicle.

3. "Anti-handling device" means a device intended to protect a mine and which is part of, linked to, attached to or placed under the mine and which activates when an attempt is made to tamper with or otherwise intentionally disturb the mine.

4. "Transfer" involves, in addition to the physical movement of anti-personnel mines into or from national territory, the transfer of title to and control over the mines, but does not involve the transfer of territory containing emplaced anti-personnel mines.

5. "Mined area" means an area which is dangerous due to the presence or suspected presence of mines.

Article 3
Exceptions

1. Notwithstanding the general obligations under Article 1, the retention or transfer of a number of anti-personnel mines for the development of and training in mine detection, mine clearance, or mine destruction techniques is permitted. The amount of such mines shall not exceed the minimum number absolutely necessary for the above-mentioned purposes.

2. The transfer of anti-personnel mines for the purpose of destruction is permitted.

Article 4
Destruction of stockpiled anti-personnel mines

Except as provided for in Article 3, each State Party undertakes to destroy or ensure the destruction of all stockpiled anti-personnel mines it owns or possesses, or that are under its jurisdiction or control, as soon as possible but not later than four years after the entry into force of this Convention for that State Party.

Article 5
Destruction of anti-personnel mines in mined areas

1. Each State Party undertakes to destroy or ensure the destruction of all anti-personnel mines in mined areas under its jurisdiction or control, as soon as possible but not later than ten years after the entry into force of this Convention for that State Party.

2. Each State Party shall make every effort to identify all areas under its jurisdiction or control in which anti-personnel mines are known or suspected to be emplaced and shall ensure as soon as possible that all anti-personnel mines in mined areas under its jurisdiction or control are perimeter-marked, monitored and protected by fencing or other means, to ensure the effective exclusion of civilians, until all anti-personnel mines contained therein have been destroyed. The marking shall at least be to the standards set out in the Protocol on Prohibitions or Restrictions on the Use of Mines, Booby-Traps and Other Devices, as amended on 3 May 1996, annexed to the Convention on Prohibitions or Restrictions on the Use of Certain Conventional Weapons Which May Be Deemed to Be Excessively Injurious or to Have Indiscriminate Effects.

3. If a State Party believes that it will be unable to destroy or ensure the destruction of all anti-personnel mines referred to in paragraph 1 within that time period, it may submit a request to a Meeting of the States Parties or a Review Conference for an extension of the deadline for completing the destruction of such anti-personnel mines, for a period of up to ten years.

4. Each request shall contain:

a) The duration of the proposed extension;

b) A detailed explanation of the reasons for the proposed extension, including:

(i) The preparation and status of work conducted under national demining programs;

(ii) The financial and technical means available to the State Party for the destruction of all the anti-personnel mines; and

(iii) Circumstances which impede the ability of the State Party to destroy all the anti-personnel mines in mined areas;

c) The humanitarian, social, economic, and environmental implications of the extension; and

d) Any other information relevant to the request for the proposed extension.

5. The Meeting of the States Parties or the Review Conference shall, taking into consideration the factors contained in

paragraph 4, assess the request and decide by a majority of votes of States Parties present and voting whether to grant the request for an extension period.

6. Such an extension may be renewed upon the submission of a new request in accordance with paragraphs 3, 4 and 5 of this Article. In requesting a further extension period a State Party shall submit relevant additional information on what has been undertaken in the previous extension period pursuant to this Article.

Article 6
International cooperation and assistance

1. In fulfilling its obligations under this Convention each State Party has the right to seek and receive assistance, where feasible, from other States Parties to the extent possible.

2. Each State Party undertakes to facilitate and shall have the right to participate in the fullest possible exchange of equipment, material and scientific and technological information concerning the implementation of this Convention. The States Parties shall not impose undue restrictions on the provision of mine clearance equipment and related technological information for humanitarian purposes.

3. Each State Party in a position to do so shall provide assistance for the care and rehabilitation, and social and economic reintegration, of mine victims and for mine awareness programs. Such assistance may be provided, inter alia, through the United Nations system, international, regional or national organizations or institutions, the International Committee of the Red Cross, national Red Cross and Red Crescent societies and their International Federation, non-governmental organizations, or on a bilateral basis.

4. Each State Party in a position to do so shall provide assistance for mine clearance and related activities. Such assistance may be provided, inter alia, through the United Nations system, international or regional organizations or institutions, non-governmental organizations or institutions, or on a bilateral basis, or by contributing to the United Nations Voluntary Trust Fund for Assistance in Mine Clearance, or other regional funds that deal with demining.

5. Each State Party in a position to do so shall provide assistance for the destruction of stockpiled anti-personnel mines.

6. Each State Party undertakes to provide information to the database on mine clearance established within the United Nations system, especially information concerning various means and technologies of mine clearance, and lists of experts, expert agencies or national points of contact on mine clearance.

7. States Parties may request the United Nations, regional organizations, other States Parties or other competent intergovernmental or non-governmental fora to assist its authorities in the elaboration of a national demining program to determine, inter alia:

a) The extent and scope of the anti-personnel mine problem;

b) The financial, technological and human resources that are required for the implementation of the program;

c) The estimated number of years necessary to destroy all anti-personnel mines in mined areas under the jurisdiction or control of the concerned State Party;

d) Mine awareness activities to reduce the incidence of mine-related injuries or deaths;

e) Assistance to mine victims;

f) The relationship between the Government of the concerned State Party and the relevant governmental, intergovernmental or non-governmental entities that will work in the implementation of the program.

8. Each State Party giving and receiving assistance under the provisions of this Article shall cooperate with a view to ensuring the full and prompt implementation of agreed assistance programs.

Article 7
Transparency measures

1. Each State Party shall report to the Secretary-General of the United Nations as soon as practicable, and in any event not later than 180 days after the entry into force of this Convention for that State Party on:

a) The national implementation measures referred to in Article 9;

b) The total of all stockpiled anti-personnel mines owned or possessed by it, or under its jurisdiction or control, to include a breakdown of the type, quantity and, if possible, lot numbers of each type of anti-personnel mine stockpiled;

c) To the extent possible, the location of all mined areas that contain, or are suspected to contain, anti-personnel mines under its jurisdiction or control, to include as much detail as possible regarding the type and quantity of each type of anti-personnel mine in each mined area and when they were emplaced;

d) The types, quantities and, if possible, lot numbers of all anti-personnel mines retained or transferred for the development of and training in mine detection, mine clearance or mine destruction techniques, or transferred for the purpose of destruction, as well as the institutions authorized by a State Party to retain or transfer anti-personnel mines, in accordance with Article 3;

e) The status of programs for the conversion or decommissioning of anti-personnel mine production facilities;

f) The status of programs for the destruction of anti-personnel mines in accordance with Articles 4 and 5, including details of the methods which will be used in destruction, the location of all destruction sites and the applicable safety and environmental standards to be observed;

g) The types and quantities of all anti-personnel mines destroyed after the entry into force of this Convention for that State Party, to include a breakdown of the quantity of each type of anti-personnel mine destroyed, in accordance with Articles 4 and 5, respectively, along with, if possible, the lot numbers of each type of anti-personnel mine in the case of destruction in accordance with Article 4;

h) The technical characteristics of each type of anti-personnel mine produced, to the extent known, and those currently owned or possessed by a State Party, giving, where reasonably possible, such categories of information as may facilitate identification and clearance of anti-personnel mines; at a minimum, this information shall include the dimensions, fusing, explosive content, metallic content, colour photographs and other information which may facilitate mine clearance; and

i) The measures taken to provide an immediate and effective warning to the population in relation to all areas identified under paragraph 2 of Article 5.

2. The information provided in accordance with this Article shall be updated by the States Parties annually, covering the last calendar year, and reported to the Secretary-General of the United Nations not later than 30 April of each year.

3. The Secretary-General of the United Nations shall transmit all such reports received to the States Parties.

Article 8
Facilitation and clarification of compliance

1. The States Parties agree to consult and cooperate with each other regarding the implementation of the provisions of this Convention, and to work together in a spirit of cooperation to facilitate compliance by States Parties with their obligations under this Convention.

2. If one or more States Parties wish to clarify and seek to resolve questions relating to compliance with the provisions of this Convention by another State Party, it may submit, through the Secretary-General of the United Nations, a Request for Clarification of that matter to that State Party. Such a request shall be accompanied by all appropriate information. Each State Party shall refrain from unfounded Requests for Clarification, care being taken to avoid abuse. A State Party that receives a Request for Clarification shall provide, through the Secretary-General of the United Nations, within 28 days to the requesting State Party all information which would assist in clarifying this matter.

3. If the requesting State Party does not receive a response through the Secretary-General of the United Nations within that time period, or deems the response to the Request for Clarification to be unsatisfactory, it may submit the matter through the Secretary-General of the United Nations to the next Meeting of the States Parties. The Secretary-General of the United Nations shall transmit the submission, accompanied by all appropriate information pertaining to the Request for Clarification, to all States Parties. All such information shall be presented to the requested State Party which shall have the right to respond.

4. Pending the convening of any meeting of the States Parties, any of the States Parties concerned may request the Secretary-General of the United Nations to exercise his or her good offices to facilitate the clarification requested.

5. The requesting State Party may propose through the Secretary-General of the United Nations the convening of a Special Meeting of the States Parties to consider the matter. The Secretary-General of the United Nations shall thereupon communicate this proposal and all information submitted by the States Parties concerned, to all States Parties with a request that they indicate whether they favour a Special Meeting of the States Parties, for the purpose of considering the matter. In the event that within 14 days from the date of such communication, at least one-third of the States Parties favours such a Special Meeting, the Secretary-General of the United Nations shall convene this Special Meeting of the States Parties within a further 14 days. A quorum for this Meeting shall consist of a majority of States Parties.

6. The Meeting of the States Parties or the Special Meeting of the States Parties, as the case may be, shall first determine whether to consider the matter further, taking into account all information submitted by the States Parties concerned. The Meeting of the States Parties or the Special Meeting of the States Parties shall make every effort to reach a decision by consensus. If despite all efforts to that end no agreement has been reached, it shall take this decision by a majority of States Parties present and voting.

7. All States Parties shall cooperate fully with the Meeting of the States Parties or the Special Meeting of the States Parties in the fulfilment of its review of the matter, including any fact-finding missions that are authorized in accordance with paragraph 8.

8. If further clarification is required, the Meeting of the States Parties or the Special Meeting of the States Parties shall authorize a fact-finding mission and decide on its mandate by a majority of States Parties present and voting. At any time the requested State Party may invite a fact-finding mission to its territory. Such a mission shall take place without a decision by a Meeting of the States Parties or a Special Meeting of the States Parties to authorize such a mission. The mission, consisting of up to 9 experts, designated and approved in accordance with paragraphs 9 and 10, may collect additional information on the spot or in other places directly related to the alleged compliance issue under the jurisdiction or control of the requested State Party.

9. The Secretary-General of the United Nations shall prepare and update a list of the names, nationalities and other relevant data of qualified experts provided by States Parties

and communicate it to all States Parties. Any expert included on this list shall be regarded as designated for all fact-finding missions unless a State Party declares its non-acceptance in writing. In the event of non-acceptance, the expert shall not participate in fact-finding missions on the territory or any other place under the jurisdiction or control of the objecting State Party, if the non-acceptance was declared prior to the appointment of the expert to such missions.

10. Upon receiving a request from the Meeting of the States Parties or a Special Meeting of the States Parties, the Secretary-General of the United Nations shall, after consultations with the requested State Party, appoint the members of the mission, including its leader. Nationals of States Parties requesting the fact-finding mission or directly affected by it shall not be appointed to the mission. The members of the fact-finding mission shall enjoy privileges and immunities under Article VI of the Convention on the Privileges and Immunities of the United Nations, adopted on 13 February 1946.

11. Upon at least 72 hours notice, the members of the fact-finding mission shall arrive in the territory of the requested State Party at the earliest opportunity. The requested State Party shall take the necessary administrative measures to receive, transport and accommodate the mission, and shall be responsible for ensuring the security of the mission to the maximum extent possible while they are on territory under its control.

12. Without prejudice to the sovereignty of the requested State Party, the fact-finding mission may bring into the territory of the requested State Party the necessary equipment which shall be used exclusively for gathering information on the alleged compliance issue. Prior to its arrival, the mission will advise the requested State Party of the equipment that it intends to utilize in the course of its fact-finding mission.

13. The requested State Party shall make all efforts to ensure that the fact-finding mission is given the opportunity to speak with all relevant persons who may be able to provide information related to the alleged compliance issue.

14. The requested State Party shall grant access for the fact-finding mission to all areas and installations under its control where facts relevant to the compliance issue could be expected to be collected. This shall be subject to any arrangements that the requested State Party considers necessary for:

a) The protection of sensitive equipment, information and areas;

b) The protection of any constitutional obligations the requested State Party may have with regard to proprietary rights, searches and seizures, or other constitutional rights; or

c) The physical protection and safety of the members of the fact-finding mission. In the event that the requested State Party makes such arrangements, it shall make every reasonable effort to demonstrate through alternative means its compliance with this Convention.

15. The fact-finding mission may remain in the territory of the State Party concerned for no more than 14 days, and at any particular site no more than 7 days, unless otherwise agreed.

16. All information provided in confidence and not related to the subject matter of the fact-finding mission shall be treated on a confidential basis.

17. The fact-finding mission shall report, through the Secretary-General of the United Nations, to the Meeting of the States Parties or the Special Meeting of the States Parties the results of its findings.

18. The Meeting of the States Parties or the Special Meeting of the States Parties shall consider all relevant information, including the report submitted by the fact-finding mission, and may request the requested State Party to take measures to address the compliance issue within a specified period of time. The requested State Party shall report on all measures taken in response to this request.

19. The Meeting of the States Parties or the Special Meeting of the States Parties may suggest to the States Parties concerned ways and means to further clarify or resolve the matter under consideration, including the initiation of appropriate procedures in conformity with international law. In circumstances where the issue at hand is determined to be due to circumstances beyond the control of the requested State Party, the Meeting of the States Parties or the Special Meeting of the States Parties may recommend appropriate measures, including the use of cooperative measures referred to in Article 6.

20. The Meeting of the States Parties or the Special Meeting of the States Parties shall make every effort to reach its decisions referred to in paragraphs 18 and 19 by consensus, otherwise by a two-thirds majority of States Parties present and voting.

Article 9
National implementation measures

Each State Party shall take all appropriate legal, administrative and other measures, including the imposition of penal sanctions, to prevent and suppress any activity prohibited to a State Party under this Convention undertaken by persons or on territory under its jurisdiction or control.

Article 10
Settlement of disputes

1. The States Parties shall consult and cooperate with each other to settle any dispute that may arise with regard to the application or the interpretation of this Convention. Each

State Party may bring any such dispute before the Meeting of the States Parties.

2. The Meeting of the States Parties may contribute to the settlement of the dispute by whatever means it deems appropriate, including offering its good offices, calling upon the States parties to a dispute to start the settlement procedure of their choice and recommending a time-limit for any agreed procedure.

3. This Article is without prejudice to the provisions of this Convention on facilitation and clarification of compliance.

Article 11
Meetings of the States Parties

1. The States Parties shall meet regularly in order to consider any matter with regard to the application or implementation of this Convention, including:

a) The operation and status of this Convention;

b) Matters arising from the reports submitted under the provisions of this Convention;

c) International cooperation and assistance in accordance with Article 6;

d) The development of technologies to clear antipersonnel mines;

e) Submissions of States Parties under Article 8; and

f) Decisions relating to submissions of States Parties as provided for in Article 5.

2. The First Meeting of the States Parties shall be convened by the Secretary-General of the United Nations within one year after the entry into force of this Convention. The subsequent meetings shall be convened by the Secretary-General of the United Nations annually until the first Review Conference.

3. Under the conditions set out in Article 8, the Secretary-General of the United Nations shall convene a Special Meeting of the States Parties.

4. States not parties to this Convention, as well as the United Nations, other relevant international organizations or institutions, regional organizations, the International Committee of the Red Cross and relevant non-governmental organizations may be invited to attend these meetings as observers in accordance with the agreed Rules of Procedure.

Article 12
Review Conferences

1. A Review Conference shall be convened by the Secretary-General of the United Nations five years after the entry into force of this Convention. Further Review Conferences shall be convened by the Secretary-General of the United Nations if so requested by one or more States Parties, provided that the interval between Review Conferences shall in no case be less than five years. All States Parties to this Convention shall be invited to each Review Conference.

2. The purpose of the Review Conference shall be:

a) To review the operation and status of this Convention;

b) To consider the need for and the interval between further Meetings of the States Parties referred to in paragraph 2 of Article 11;

c) To take decisions on submissions of States Parties as provided for in Article 5; and

d) To adopt, if necessary, in its final report conclusions related to the implementation of this Convention.

3. States not parties to this Convention, as well as the United Nations, other relevant international organizations or institutions, regional organizations, the International Committee of the Red Cross and relevant non-governmental organizations may be invited to attend each Review Conference as observers in accordance with the agreed Rules of Procedure.

Article 13
Amendments

1. At any time after the entry into force of this Convention any State Party may propose amendments to this Convention. Any proposal for an amendment shall be communicated to the Depositary, who shall circulate it to all States Parties and shall seek their views on whether an Amendment Conference should be convened to consider the proposal. If a majority of the States Parties notify the Depositary no later than 30 days after its circulation that they support further consideration of the proposal, the Depositary shall convene an Amendment Conference to which all States Parties shall be invited.

2. States not parties to this Convention, as well as the United Nations, other relevant international organizations or institutions, regional organizations, the International Committee of the Red Cross and relevant non-governmental organizations may be invited to attend each Amendment Conference as observers in accordance with the agreed Rules of Procedure.

3. The Amendment Conference shall be held immediately following a Meeting of the States Parties or a Review Conference unless a majority of the States Parties request that it be held earlier.

4. Any amendment to this Convention shall be adopted by a majority of two-thirds of the States Parties present and voting at the Amendment Conference. The Depositary shall communicate any amendment so adopted to the States Parties.

5. An amendment to this Convention shall enter into force for all States Parties to this Convention which have accepted

it, upon the deposit with the Depositary of instruments of acceptance by a majority of States Parties. Thereafter it shall enter into force for any remaining State Party on the date of deposit of its instrument of acceptance.

Article 14
Costs

1. The costs of the Meetings of the States Parties, the Special Meetings of the States Parties, the Review Conferences and the Amendment Conferences shall be borne by the States Parties and States not parties to this Convention participating therein, in accordance with the United Nations scale of assessment adjusted appropriately.

2. The costs incurred by the Secretary-General of the United Nations under Articles 7 and 8 and the costs of any fact-finding mission shall be borne by the States Parties in accordance with the United Nations scale of assessment adjusted appropriately.

Article 15
Signature

This Convention, done at Oslo, Norway, on 18 September 1997, shall be open for signature at Ottawa, Canada, by all States from 3 December 1997 until 4 December 1997, and at the United Nations Headquarters in New York from 5 December 1997 until its entry into force.

Article 16
Ratification, acceptance, approval or accession

1. This Convention is subject to ratification, acceptance or approval of the Signatories.

2. It shall be open for accession by any State which has not signed the Convention.

3. The instruments of ratification, acceptance, approval or accession shall be deposited with the Depositary.

Article 17
Entry into force

1. This Convention shall enter into force on the first day of the sixth month after the month in which the 40th instrument of ratification, acceptance, approval or accession has been deposited.

2. For any State which deposits its instrument of ratification, acceptance, approval or accession after the date of the deposit of the 40th instrument of ratification, acceptance, approval or accession, this Convention shall enter into force on the first day of the sixth month after the date on which that State has deposited its instrument of ratification, acceptance, approval or accession.

Article 18
Provisional application

Any State may at the time of its ratification, acceptance, approval or accession, declare that it will apply provisionally paragraph 1 of Article 1 of this Convention pending its entry into force.

Article 19
Reservations

The Articles of this Convention shall not be subject to reservations.

Article 20
Duration and withdrawal

1. This Convention shall be of unlimited duration.

2. Each State Party shall, in exercising its national sovereignty, have the right to withdraw from this Convention. It shall give notice of such withdrawal to all other States Parties, to the Depositary and to the United Nations Security Council. Such instrument of withdrawal shall include a full explanation of the reasons motivating this withdrawal.

3. Such withdrawal shall only take effect six months after the receipt of the instrument of withdrawal by the Depositary. If, however, on the expiry of that six- month period, the withdrawing State Party is engaged in an armed conflict, the withdrawal shall not take effect before the end of the armed conflict.

4. The withdrawal of a State Party from this Convention shall not in any way affect the duty of States to continue fulfilling the obligations assumed under any relevant rules of international law.

Article 21
Depositary

The Secretary-General of the United Nations is hereby designated as the Depositary of this Convention.

Article 22
Authentic texts

The original of this Convention, of which the Arabic, Chinese, English, French, Russian and Spanish texts are equally authentic, shall be deposited with the Secretary-General of the United Nations.

AMENDED PROTOCOL II OF THE CONVENTION ON PROHIBITIONS OR RESTRICTIONS ON THE USE OF CERTAIN CONVENTIONAL WEAPONS WHICH MAY BE DEEMED TO BE EXCESSIVELY INJURIOUS OR TO HAVE INDISCRIMINATE EFFECTS (CONVENTION ON CONVENTIONAL WEAPONS [CCW])

Article I - Scope of application

1. This Protocol relates to the use on land of the mines, booby-traps and other devices, defined herein, including mines laid to interdict beaches, waterway crossings or river crossings, but does not apply to the use of anti-ship mines at sea or in inland waterways.

2. This Protocol shall apply, in addition to situations referred to in Article I of this Convention, to situations referred to in Article 3 common to the Geneva Conventions of 12 August 1949. This Protocol shall not apply to situations of internal disturbances and tensions, such as riots, isolated and sporadic acts of violence and other acts of a similar nature, as not being armed conflicts.

3. In case of armed conflicts not of an international character occurring in the territory of one of the High Contracting Parties, each party to the conflict shall be bound to apply the prohibitions and restrictions of this Protocol.

4. Nothing in this Protocol shall be invoked for the purpose of affecting the sovereignty of a State or the responsibility of the Government, by all legitimate means, to maintain or re-establish law and order in the State or to defend the national unity and territorial integrity of the State.

5. Nothing in this Protocol shall be invoked as a justification for intervening, directly or indirectly, for any reason whatever, in the armed conflict or in the internal or external affairs of the High Contracting Party in the territory of which that conflict occurs.

6. The application of the provisions of this Protocol to parties to a conflict, which are not High Contracting Parties that have accepted this Protocol, shall not change their legal status or the legal status of a disputed territory, either explicitly or implicitly.

Article 2 - Definitions

For the purpose of this Protocol:

1. "Mine" means a munition placed under, on or near the ground or other surface area and designed to be exploded by the presence, proximity or contact of a person or vehicle.

2. "Remotely-delivered mine" means a mine not directly emplaced but delivered by artillery, missile, rocket, mortar, or similar means, or dropped from an aircraft. Mines delivered from a land-based system from less than 500 metres are not considered to be "remotely delivered", provided that they are used in accordance with Article 5 and other relevant Articles of this Protocol.

3. "Anti-personnel mine" means a mine primarily designed to be exploded by the presence, proximity or contact of a person and that will incapacitate, injure or kill one or more persons.

4. "Booby-trap" means any device or material which is designed, constructed or adapted to kill or injure, and which functions unexpectedly when a person disturbs or approaches an apparently harmless object or performs an apparently safe act.

5. "Other devices" means manually-emplaced munitions and devices including improvised explosive devices designed to kill, injure or damage and which are actuated manually, by remote control or automatically after a lapse of time.

6. "Military objective" means, so far as objects are concerned, any object which by its nature, location, purpose or use makes an effective contribution to military action and whose total or partial destruction, capture or neutralization, in the circumstances ruling at the time, offers a definite military advantage.

7. "Civilian objects" are all objects which are not military objectives as defined in paragraph 6 of this Article.

8. "Minefield" is a defined area in which mines have been emplaced and "mined area" is an area which is dangerous due to the presence of mines. "Phoney minefield" means an area free of mines that simulates a minefield. The term "minefield" includes phoney minefields.

9. "Recording" means a physical, administrative and technical operation designed to obtain, for the purpose of registration in official records, all available information facilitating the location of minefields, mined areas, mines, booby-traps and other devices.

10. "Self-destruction mechanism" means an incorporated or externally attached automatically-functioning mechanism which secures the destruction of the munition into which it is incorporated or to which it is attached.

11. "Self-neutralization mechanism" means an incorporated automatically-functioning mechanism which renders inoperable the munition into which it is incorporated.

12. "Self-deactivating" means automatically rendering a munition inoperable by means of the irreversible exhaustion of a component, for example, a battery, that is essential to the operation of the munition.

13. "Remote control" means control by commands from a distance.

14. "Anti-handling device" means a device intended to protect a mine and which is part of, linked to, attached to or

placed under the mine and which activates when an attempt is made to tamper with the mine.

15. "Transfer" involves, in addition to the physical movement of mines into or from national territory, the transfer of title to and control over the mines, but does not involve the transfer of territory containing emplaced mines.

Article 3 - General restrictions on the use of mines, booby-traps and other devices

1. This Article applies to:
   (a) mines;
   (b) booby-traps; and
   (c) other devices.

2. Each High Contracting Party or party to a conflict is, in accordance with the provisions of this Protocol, responsible for all mines, booby-traps, and other devices employed by it and undertakes to clear, remove, destroy or maintain them as specified in Article 10 of this Protocol.

3. It is prohibited in all circumstances to use any mine, booby-trap or other device which is designed or of a nature to cause superfluous injury or unnecessary suffering.

4. Weapons to which this Article applies shall strictly comply with the standards and limitations specified in the Technical Annex with respect to each particular category.

5. It is prohibited to use mines, booby-traps or other devices which employ a mechanism or device specifically designed to detonate the munition by the presence of commonly available mine detectors as a result of their magnetic or other non-contact influence during normal use in detection operations.

6. It is prohibited to use a self-deactivating mine equipped with an anti-handling device that is designed in such a manner that the anti-handling device is capable of functioning after the mine has ceased to be capable of functioning.

7. It is prohibited in all circumstances to direct weapons to which this Article applies, either in offence, defence or by way of reprisals, against the civilian population as such or against individual civilians or civilian objects.

8. The indiscriminate use of weapons to which this Article applies is prohibited. Indiscriminate use is any placement of such weapons:

(a) which is not on, or directed against, a military objective. In case of doubt as to whether an object which is normally dedicated to civilian purposes, such as a place of worship, a house or other dwelling or a school, is being used to make an effective contribution to military action, it shall be presumed not to be so used; or

(b) which employs a method or means of delivery which cannot be directed at a specific military objective; or

(c) which may be expected to cause incidental loss of civilian life, injury to civilians, damage to civilian objects, or a combination thereof, which would be excessive in relation to the concrete and direct military advantage anticipated.

9. Several clearly separated and distinct military objectives located in a city, town, village or other area containing a similar concentration of civilians or civilian objects are not to be treated as a single military objective.

10. All feasible precautions shall be taken to protect civilians from the effects of weapons to which this Article applies. Feasible precautions are those precautions which are practicable or practically possible taking into account all circumstances ruling at the time, including humanitarian and military considerations. These circumstances include, but are not limited to:

(a) the short- and long-term effect of mines upon the local civilian population for the duration of the minefield;

(b) possible measures to protect civilians (for example, fencing, signs, warning and monitoring);

(c) the availability and feasibility of using alternatives; and

(d) the short- and long-term military requirements for a minefield.

11. Effective advance warning shall be given of any emplacement of mines, booby-traps and other devices which may affect the civilian population, unless circumstances do not permit.

Article 4 - Restrictions on the use of anti-personnel mines

It is prohibited to use anti-personnel mines which are not detectable, as specified in paragraph 2 of the Technical Annex.

Article 5 - Restrictions on the use of anti-personnel mines other than remotely-delivered mines

1. This Article applies to anti-personnel mines other than remotely-delivered mines.

2. It is prohibited to use weapons to which this Article applies which are not in compliance with the provisions on self-destruction and self-deactivation in the Technical Annex, unless:

(a) such weapons are placed within a perimeter-marked area which is monitored by military personnel and protected by fencing or other means, to ensure the effective exclusion of civilians from the area. The marking must be of a distinct and durable character and must at least be visible to a person who is about to enter the perimeter-marked area; and

(b) such weapons are cleared before the area is abandoned, unless the area is turned over to the forces of another State which accept responsibility for the maintenance of the protections required by this Article and the subsequent clearance of those weapons.

3. A party to a conflict is relieved from further compliance with the provisions of sub-paragraphs 2 (a) and 2 (b) of this Article only if such compliance is not feasible due to forcible loss of control of the area as a result of enemy military action, including situations where direct enemy military action makes it impossible to comply. If that party regains control of the area, it shall resume compliance with the provisions of sub-paragraphs 2 (a) and 2 (b) of this Article.

4. If the forces of a party to a conflict gain control of an area in which weapons to which this Article applies have been laid, such forces shall, to the maximum extent feasible, maintain and, if necessary, establish the protections required by this Article until such weapons have been cleared.

5. All feasible measures shall be taken to prevent the unauthorized removal, defacement, destruction or concealment of any device, system or material used to establish the perimeter of a perimeter-marked area.

6. Weapons to which this Article applies which propel fragments in a horizontal arc of less than 90 degrees and which are placed on or above the ground may be used without the measures provided for in sub-paragraph 2 (a) of this Article for a maximum period of 72 hours, if:

(a) they are located in immediate proximity to the military unit that emplaced them; and

(b) the area is monitored by military personnel to ensure the effective exclusion of civilians.

Article 6 - Restrictions on the use of remotely-delivered mines

1. It is prohibited to use remotely-delivered mines unless they are recorded in accordance with sub-paragraph I (b) of the Technical Annex.

2. It is prohibited to use remotely-delivered anti-personnel mines which are not in compliance with the provisions on self-destruction and self-deactivation in the Technical Annex.

3. It is prohibited to use remotely-delivered mines other than anti-personnel mines, unless, to the extent feasible, they are equipped with an effective self-destruction or self-neutralization mechanism and have a back-up self-deactivation feature, which is designed so that the mine will no longer function as a mine when the mine no longer serves the military purpose for which it was placed in position.

4. Effective advance warning shall be given of any delivery or dropping of remotely-delivered mines which may affect the civilian population, unless circumstances do not permit.

Article 7 - Prohibitions on the use of booby-traps and other devices

1. Without prejudice to the rules of international law applicable in armed conflict relating to treachery and perfidy, it is prohibited in all circumstances to use booby-traps and other devices which are in any way attached to or associated with:

(a) internationally recognized protective emblems, signs or signals;

(b) sick, wounded or dead persons;

(c) burial or cremation sites or graves;

(d) medical facilities, medical equipment, medical supplies or medical transportation;

(e) children's toys or other portable objects or products specially designed for the feeding, health, hygiene, clothing or education of children;

(f) food or drink;

(g) kitchen utensils or appliances except in military establishments, military locations or military supply depots;

(h) objects clearly of a religious nature;

(i) historic monuments, works of art or places of worship which constitute the cultural or spiritual heritage of peoples; or

(j) animals or their carcasses.

2. It is prohibited to use booby-traps or other devices in the form of apparently harmless portable objects which are specifically designed and constructed to contain explosive material.

3. Without prejudice to the provisions of Article 3, it is prohibited to use weapons to which this Article applies in any city, town, village or other area containing a similar concentration of civilians in which combat between ground forces is not taking place or does not appear to be imminent, unless either:

(a) they are placed on or in the close vicinity of a military objective; or

(b) measures are taken to protect civilians from their effects, for example, the posting of warning sentries, the issuing of warnings or the provision of fences.

Article 8 - Transfers

1. In order to promote the purposes of this Protocol, each High Contracting Party:

(a) undertakes not to transfer any mine the use of which is prohibited by this Protocol;

(b) undertakes not to transfer any mine to any recipient other than a State or a State agency authorized to receive such transfers;

(c) undertakes to exercise restraint in the transfer of any mine the use of which is restricted by this Protocol. In particular, each High Contracting Party undertakes not to transfer any anti-personnel mines to States which are not bound by this Protocol, unless the recipient State agrees to apply this Protocol; and

(d) undertakes to ensure that any transfer in accordance with this Article takes place in full compliance, by both the transferring and the recipient State, with the relevant provisions of this Protocol and the applicable norms of international humanitarian law.

2. In the event that a High Contracting Party declares that it will defer compliance with specific provisions on the use of certain mines, as provided for in the Technical Annex, sub-paragraph I (a) of this Article shall however apply to such mines.

3. All High Contracting Parties, pending the entry into force of this Protocol, will refrain from any actions which would be inconsistent with sub-paragraph I (a) of this Article.

Article 9 - Recording and use of information on minefields, mined areas, mines, booby-traps and other devices

1. All information concerning minefields, mined areas, mines, booby-traps and other devices shall be recorded in accordance with the provisions of the Technical Annex.

2. All such records shall be retained by the parties to a conflict, who shall, without delay after the cessation of active hostilities, take all necessary and appropriate measures, including the use of such information, to protect civilians from the effects of minefields, mined areas, mines, booby-traps and other devices in areas under their control. At the same time, they shall also make available to the other party or parties to the conflict and to the Secretary-General of the United Nations all such information in their possession concerning minefields, mined areas, mines, booby-traps and other devices laid by them in areas no longer under their control; provided, however, subject to reciprocity, where the forces of a party to a conflict are in the territory of an adverse party, either party may withhold such information from the Secretary General and the other party, to the extent that security interests require such withholding, until neither party is in the territory of the other. In the latter case, the information withheld shall be disclosed as soon as those security interests permit. Wherever possible, the parties to the conflict shall seek, by mutual agreement, to provide for the release of such information at the earliest possible time in a manner consistent with the security interests of each party.

3. This Article is without prejudice to the provisions of Articles 10 and 12 of this Protocol.

Article 10 - Removal of minefields, mined areas, mines, booby-traps and other devices and international cooperation

1. Without delay after the cessation of active hostilities, all minefields, mined areas, mines, booby-traps and other devices shall be cleared, removed, destroyed or maintained in accordance with Article 3 and paragraph 2 of Article 5 of this Protocol.

2. High Contracting Parties and parties to a conflict bear such responsibility with respect to minefields, mined areas, mines, booby-traps and other devices in areas under their control.

3. With respect to minefields, mined areas, mines, booby-traps and other devices laid by a party in areas over which it no longer exercises control, such party shall provide to the party in control of the area pursuant to paragraph 2 of this Article, to the extent permitted by such party, technical and material assistance necessary to fulfil such responsibility.

4. At all times necessary, the parties shall endeavour to reach agreement, both among themselves and, where appropriate, with other States and with international organizations, on the provision of technical and material assistance, including, in appropriate circumstances, the undertaking of joint operations necessary to fulfil such responsibilities.

Article 11 - Technological cooperation and assistance

1. Each High Contracting Party undertakes to facilitate and shall have the right to participate in the fullest possible exchange of equipment, material and scientific and technological information concerning the implementation of this Protocol and means of mine clearance. In particular, High Contracting Parties shall not impose undue restrictions on the provision of mine clearance equipment and related technological information for humanitarian purposes.

2. Each High Contracting Party undertakes to provide information to the database on mine clearance established within the United Nations System, especially information concerning various means and technologies of mine clearance, and lists of experts, expert agencies or national points of contact on mine clearance.

3. Each High Contracting Party in a position to do so shall provide assistance for mine clearance through the United Nations System, other international bodies or on a bilateral basis, or contribute to the United Nations Voluntary Trust Fund for Assistance in Mine Clearance.

4. Requests by High Contracting Parties for assistance, substantiated by relevant information, may be submitted to the United Nations, to other appropriate bodies or to other States. These requests may be submitted to the Secretary-General of the United Nations, who shall transmit them to all High Contracting Parties and to relevant international organizations.

5. In the case of requests to the United Nations, the Secretary-General of the United Nations, within the resources available to the Secretary-General of the United Nations, may take appropriate steps to assess the situation and, in cooperation with the requesting High Contracting Party, determine the appropriate provision of assistance in mine clearance or implementation of the Protocol. The

Secretary-General may also report to High Contracting Parties on any such assessment as well as on the type and scope of assistance required.

6. Without prejudice to their constitutional and other legal provisions, the High Contracting Parties undertake to cooperate and transfer technology to facilitate the implementation of the relevant prohibitions and restrictions set out in this Protocol.

7. Each High Contracting Party has the right to seek and receive technical assistance, where appropriate, from another High Contracting Party on specific relevant technology, other than weapons technology, as necessary and feasible, with a view to reducing any period of deferral for which provision is made in the Technical Annex.

Article 12 - Protection from the effects of minefields, mined areas, mines, booby-traps and other devices

1. Application

a) With the exception of the forces and missions referred to in sub-paragraph 2(a) (i) of this Article, this Article applies only to missions which are performing functions in an area with the consent of the High Contracting Party on whose territory the functions are performed.

(b) The application of the provisions of this Article to parties to a conflict which are not High Contracting Parties shall not change their legal status or the legal status of a disputed territory, either explicitly or implicitly.

(c) The provisions of this Article are without prejudice to existing international humanitarian law, or other international instruments as applicable, or decisions by the Security Council of the United Nations, which provide for a higher level of protection to personnel functioning in accordance with this Article.

2. Peace-keeping and certain other forces and missions

(a) This paragraph applies to:

(i) any United Nations force or mission performing peace-keeping, observation or similar functions in any area in accordance with the Charter of the United Nations;

(ii) any mission established pursuant to Chapter VIII of the Charter of the United Nations and performing its functions in the area of a conflict.

(b) Each High Contracting Party or party to a conflict, if so requested by the head of a force or mission to which this paragraph applies, shall:

(i) so far as it is able, take such measures as are necessary to protect the force or mission from the effects of mines, booby-traps and other devices in any area under its control;

(ii) if necessary in order effectively to protect such personnel, remove or render harmless, so far as it is able, all mines, booby-traps and other devices in that area; and

(iii) inform the head of the force or mission of the location of all known minefields, mined areas, mines, booby-traps and other devices in the area in which the force or mission is performing its functions and, so far as is feasible, make available to the head of the force or mission all information in its possession concerning such minefields, mined areas, mines, booby-traps and other devices.

3. Humanitarian and fact-finding missions of the United Nations System

(a) This paragraph applies to any humanitarian or fact-finding mission of the United Nations System.

(b) Each High Contracting Party or party to a conflict, if so requested by the head of a mission to which this paragraph applies, shall:

(i) provide the personnel of the mission with the protections set out in sub-paragraph 2(b) (i) of this Article; and

(ii) if access to or through any place under its control is necessary for the performance of the mission's functions and in order to provide the personnel of the mission with safe passage to or through that place:

(aa) unless on-going hostilities prevent, inform the head of the mission of a safe route to that place if such information is available; or

(bb) if information identifying a safe route is not provided in accordance with sub-paragraph (aa), so far as is necessary and feasible, clear a lane through minefields.

4. Missions of the International Committee of the Red Cross

(a) This paragraph applies to any mission of the International Committee of the Red Cross performing functions with the consent of the host State or States as provided for by the Geneva Conventions of 12 August 1949 and, where applicable, their Additional Protocols.

(b) Each High Contracting Party or party to a conflict, if so requested by the head of a mission to which this paragraph applies, shall:

(i) provide the personnel of the mission with the protections set out in sub-paragraph 2(b) (i) of this Article; and

(ii) take the measures set out in sub-paragraph 3(b) (ii) of this Article.

5. Other humanitarian missions and missions of enquiry

(a) Insofar as paragraphs 2, 3 and 4 above do not apply to them, this paragraph applies to the following missions when they are performing functions in the area of a conflict or to assist the victims of a conflict:

(i) any humanitarian mission of a national Red Cross or Red Crescent Society or of their International Federation;

(ii) any mission of an impartial humanitarian organization, including any impartial humanitarian demining mission; and

(iii) any mission of enquiry established pursuant to the provisions of the Geneva Conventions of 12 August 1949 and, where applicable, their Additional Protocols.

(b) Each High Contracting Party or party to a conflict, if so requested by the head of a mission to which this paragraph applies, shall, so far as is feasible:

(i) provide the personnel of the mission with the protections set out in sub-paragraph 2(b) (i) of this Article, and

(ii) take the measures set out in sub-paragraph 3(b) (ii) of this Article.

6. Confidentiality. All information provided in confidence pursuant to this Article shall be treated by the recipient in strict confidence and shall not be released outside the force or mission concerned without the express authorization of the provider of the information.

7. Respect for laws and regulations. Without prejudice to such privileges and immunities as they may enjoy or to the requirements of their duties, personnel participating in the forces and missions referred to in this Article shall:

(a) respect the laws and regulations of the host State; and

(b) refrain from any action or activity incompatible with the impartial and international nature of their duties.

Article 13 - Consultations of High Contracting Parties

1. The High Contracting Parties undertake to consult and cooperate with each other on all issues related to the operation of this Protocol. For this purpose, a conference of High Contracting Parties shall be held annually.

2. Participation in the annual conferences shall be determined by their agreed Rules of Procedure.

3. The work of the conference shall include:

(a) review of the operation and status of this Protocol;

(b) consideration of matters arising from reports by High Contracting Parties according to paragraph 4 of this Article;

(c) preparation for review conferences; and

(d) consideration of the development of technologies to protect civilians against indiscriminate effects of mines.

4. The High Contracting Parties shall provide annual reports to the Depositary, who shall circulate them to all High Contracting Parties in advance of the Conference, on any of the following matters:

(a) dissemination of information on this Protocol to their armed forces and to the civilian population;

(b) mine clearance and rehabilitation programmes;

(c) steps taken to meet technical requirements of this Protocol and any other relevant information pertaining thereto;

(d) legislation related to this Protocol;

(e) measures taken on international technical information exchange, on international cooperation on mine clearance, and on technical cooperation and assistance; and

(f) other relevant matters.

5. The cost of the Conference of High Contracting Parties shall be borne by the High Contracting Parties and States not parties participating in the work of the Conference, in accordance with the United Nations scale of assessment adjusted appropriately.

Article 14 - Compliance

1. Each High Contracting Party shall take all appropriate steps, including legislative and other measures, to prevent and suppress violations of this Protocol by persons or on territory under its jurisdiction or control.

2. The measures envisaged in paragraph I of this Article include appropriate measures to ensure the imposition of penal sanctions against persons who, in relation to an armed conflict and contrary to the provisions of this Protocol, wilfully kill or cause serious injury to civilians and to bring such persons to justice.

3. Each High Contracting Party shall also require that its armed forces issue relevant military instructions and operating procedures and that armed forces personnel receive training commensurate with their duties and responsibilities to comply with the provisions of this Protocol.

4. The High Contracting Parties undertake to consult each other and to cooperate with each other bilaterally, through the Secretary-General of the United Nations or through other appropriate international procedures, to resolve any problems that may arise with regard to the interpretation and application of the provisions of this Protocol.

Technical Annex

1. Recording

(a) Recording of the location of mines other than remotely-delivered mines, minefields, mined areas, booby-traps and other devices shall be carried out in accordance with the following provisions:

(i) the location of the minefields, mined areas and areas of booby-traps and other devices shall be specified accurately by relation to the coordinates of at least two reference points and the estimated dimensions of the area containing these weapons in relation to those reference points;

(ii) maps, diagrams or other records shall be made in such a way as to indicate the location of minefields, mined areas, booby-traps and other devices in relation to reference points, and these records shall also indicate their perimeters and extent;

(iii) for purposes of detection and clearance of mines, booby-traps and other devices, maps, diagrams or other records shall contain complete information on the type, number, emplacing method, type of fuse and life time, date and time of laying, anti-handling devices (if any) and other relevant information on all these weapons laid. Whenever feasible the minefield record shall show the exact location of every mine, except in row minefields where the row location is sufficient. The precise location and operating mechanism of each booby-trap laid shall be individually recorded.

(b) The estimated location and area of remotely-delivered mines shall be specified by coordinates of reference points (normally corner points) and shall be ascertained and when feasible marked on the ground at the earliest opportunity. The total number and types of mines laid, the date and time of laying and the self-destruction time periods shall also be recorded.

(c) Copies of records shall be held at a level of command sufficient to guarantee their safety as far as possible.

(d) The use of mines produced after the entry into force of this Protocol is prohibited unless they are marked in English or in the respective national language or languages with the following information:

(i) name of the country of origin;

(ii) month and year of production; and

(iii) serial number or lot number.

The marking should be visible, legible, durable and resistant to environmental effects, as far as possible.

2. Specifications on detectability

(a) With respect to anti-personnel mines produced after 1 January 1997, such mines shall incorporate in their construction a material or device that enables the mine to be detected by commonly-available technical mine detection equipment and provides a response signal equivalent to a signal from 8 grammes or more of iron in a single coherent mass.

(b) With respect to anti-personnel mines produced before 1 January 1997, such mines shall either incorporate in their construction, or have attached prior to their emplacement, in a manner not easily removable, a material or device that enables the mine to be detected by commonly-available technical mine detection equipment and provides a response signal equivalent to a signal from 8 grammes or more of iron in a single coherent mass.

(c) In the event that a High Contracting Party determines that it cannot immediately comply with sub-paragraph (b), it may declare at the time of its notification of consent to be bound by this Protocol that it will defer compliance with sub-paragraph (b) for a period not to exceed 9 years from the entry into force of this Protocol. In the meantime it shall, to the extent feasible, minimize the use of anti-personnel mines that do not so comply.

3. Specifications on self-destruction and self-deactivation

(a) All remotely-delivered anti-personnel mines shall be designed and constructed so that no more than 10% of activated mines will fail to self-destruct within 30 days after emplacement, and each mine shall have a back-up self-deactivation feature designed and constructed so that, in combination with the self-destruction mechanism, no more than one in one thousand activated mines will function as a mine 120 days after emplacement.

(b) All non-remotely delivered anti-personnel mines, used outside marked areas, as defined in Article 5 of this Protocol, shall comply with the requirements for self-destruction and self-deactivation stated in sub-paragraph (a).

(c) In the event that a High Contracting Party determines that it cannot immediately comply with sub-paragraphs (a) and/or (b), it may declare at the time of its notification of consent to be bound by this Protocol, that it will, with respect to mines produced prior to the entry into force of this Protocol defer compliance with sub-paragraphs (a) and/or (b) for a period not to exceed 9 years from the entry into force of this Protocol. During this period of deferral, the High Contracting Party shall:

(i) undertake to minimize, to the extent feasible, the use of anti-personnel mines that do not so comply, and

(ii) with respect to remotely-delivered anti-personnel mines, comply with either the requirements for self-destruction or the requirements for self-deactivation and, with respect to other anti-personnel mines comply with at least the requirements for self-deactivation.

4. International signs for minefields and mined areas. Signs similar to the example attached [1] and as specified below shall be utilized in the marking of minefields and mined areas to ensure their visibility and recognition by the civilian population:

(a) size and shape: a triangle or square no smaller than 28 centimetres (11 inches) by 20 centimetres (7.9 inches) for a triangle, and 15 centimetres (6 inches) per side for a square;

(b) colour: red or orange with a yellow reflecting border.

Appendix F

Signatories to the Ottawa Convention and Their Alternatives to Landmines

The *Convention on the Prohibition of the Use, Stockpiling, Production and Transfer of Anti-Personnel Mines and on Their Destruction* (known as the Ottawa Convention) was open for signature from December 3, 1997, until its entry into force on March 1, 1999, six months after it had been ratified, accepted, approved, or acceded to by 40 countries. After that date, no country was allowed to sign it and ratify it later. Countries could join (become a party to) the treaty, however, through a one-step procedure known as accession.

As of September 2000, 107 countries had ratified, accepted, approved or acceded to the convention. Although few of these countries are actively searching for or developing alternatives to landmines, many are monitoring international developments in this area; several countries are participating in a North Atlantic Treaty Organization (NATO) study on the consequences of the APL ban and possible technological alternatives that do not have the negative effects of APL. The Committee on Alternative Technologies to Replace Antipersonnel Landmines found a few instances of countries other than the United States identifying or working to identify alternatives to APL.

AUSTRALIA

An Australian company, Metal Storm, has developed an all-electronic firing system that represents a breakthrough in gun technology, which the company believes could lead to "the development of an area denial weapons system to replace antipersonnel landmines" (Metal Storm, 2000a). The Australian Army has approved a three-year program for the development of a prototype minefield-replacement mortar-box system, utilizing Metal Storm technology (Metal Storm, 2000b). Conceptually, this application would be similar in operation to the U.S. Claymore and the French Sphinx-Moder. According to the company's description, a man-in-the-loop, after observing and identifying a target, would fire a launcher sending a variety of projectiles into the protected area.

CANADA

The Canadian Centre for Mine-Action Technologies (CCMAT), a joint initiative of the Department of National Defence and Industry Canada, is mandated to "conduct research and gather information to show that viable and more humane alternatives [to APL], which do not target civilians, can be developed." CCMAT also conducts research on demining technologies, medical treatment, and the rehabilitation of mine victims (CDND, 1998). CCMAT is exploring nonlethal alternatives only (ICBL, 2000). It is also conducting a series of studies "to determine the impact of removing antipersonnel landmines on land force operations and to determine if replacement technologies are necessary" (Roy and Friesen, 1999). The first volume in this series, a study on the historical uses of APL, was made available to the Committee on Alternative Technologies to Replace Antipersonnel Landmines.

According to the Antipersonnel Mine Operational Planning and Policy Guidelines for the Canadian Forces, Canada would replace its APL with "a mix of sensors, command-detonated weapons [such as the M-18 Claymore reclassified as C19s], additional infantry, artillery, armour and air-delivered weapons" (Fredenburg, 1997).

FRANCE

The Sphinx-Moder (described in Chapter 6) is designed to fire wounding, warning, or practice munitions. It is being produced in series and has been adopted by the French Army to take the place of antipersonnel mines.

JAPAN

The Japanese Defense Agency is developing an alternative weapon system to APL called the "antipersonnel obstacle system," which combines sensors and remote control.

The system is detonated manually. Until a new system is developed, the Defense Agency will use "directional-multiple-shots" as an alternative weapon (ICBL, 1999).

RUSSIA

Although Russia is not a party to the Ottawa Convention, it has been focusing its efforts on the research and development of landmine alternatives (ICBL, 2000). Few details are available, but researchers appear to be improving antipersonnel munitions that would not be considered APL under the convention, including munitions actuated by an operator by radio, wire, or automatically after a definite period of time. Research on alternatives is being conducted by the State Research and Development Engineer Institute and the Science-Research Machinery Building Institute (ICBL, 1999).

SWITZERLAND

Switzerland conducted an investigation of nonlethal APL alternatives, video monitors, and various technical sensors. However, the program did not lead to feasible results and has been terminated (ICBL, 2000).

UNITED KINGDOM

The U.K. Ministry of Defence is investigating possible nonlethal alternatives, but it is not yet known whether these will be produced (ICBL, 2000).

The following 107 countries had ratified or acceded to the Ottawa Convention as of September 2000:

Albania
Andorra
Antigua and Barbuda
Argentina
Australia
Austria
Bahamas
Bangladesh
Barbados
Belgium
Belize
Benin
Bolivia
Bosnia and Herzegovina
Botswana
Brazil
Bulgaria
Burkina Faso
Cambodia
Canada
Chad
Colombia
Costa Rica
Côte d'Ivoire
Croatia
Czech Republic
Denmark
Djibouti
Dominica
Dominican Republic
Ecuador
El Salvador
Equatorial Guinea
Fiji
France
Gabon
Germany
Ghana
Grenada
Guatemala
Guinea
Holy See
Honduras
Hungary
Iceland
Ireland
Italy
Jamaica
Japan
Jordan
Kiribati
Lesotho
Liberia
Liechtenstein
Luxembourg
Macedonia, FYR
Madagascar
Maldives
Malaysia
Malawi
Mali
Mauritania
Mauritius
Mexico
Moldova
Monaco
Mozambique
Nauru
Namibia
Netherlands
New Zealand
Nicaragua
Niger
Niue
Norway
Panama
Paraguay
Peru
Philippines
Portugal
Qatar
Rwanda
Saint Kitts and Nevis
Saint Lucia
Samoa
San Marino
Senegal
Seychelles
Slovakia
Slovenia
Solomon Islands
South Africa
Spain
Swaziland
Sweden
Switzerland
Tajikistan
Thailand
Togo
Trinidad and Tobago
Tunisia
Turkmenistan
Uganda
United Kingdom
Venezuela
Yemen
Zimbabwe

Between September 2000 and the publication of this report, two countries ratified the convention:

Romania
United Republic of Tanzania

The following 30 countries have signed the convention but have not ratified it and, therefore, are not yet parties to the convention:

Algeria	Chile	Guinea-Bissau	Malta	Sudan
Angola	Cook Islands	Guyana	Marshall Islands	Suriname
Brunei Darussalam	Cyprus	Haiti	Poland	Ukraine
Burundi	Ethiopia	Indonesia	Saint Vincent and the Grenadines	Uruguay
Cameroon	Gambia	Kenya	São Tomé e Principe	Vanuatu
Cape Verde	Greece	Lithuania	Sierra Leone	Zambia

The following countries had not signed the convention as of December 2000:

Afghanistan	D.R. Congo	Korea, North	Myanmar (Burma)	Sri Lanka
Armenia	Egypt	Korea, South	Nepal	Syria
Azerbaijan	Eritrea	Kuwait	Nigeria	Tonga
Bahrain	Estonia	Kyrgyzstan	Oman	Turkey
Belarus	Finland	Laos	Pakistan	Tuvalu
Bhutan	Georgia	Latvia	Palau	United Arab Emirates
Central African Republic	India	Lebanon	Papua New Guinea	United States of America
China	Iran	Libya	Russia	Uzbekistan
Comoros	Iraq	Micronesia	Saudi Arabia	Vietnam
Congo (Brazzaville)	Israel	Mongolia	Singapore	Yugoslavia
Cuba	Kazakhstan	Morocco	Somalia	

REFERENCES

CDND (Canadian Department of National Defence). 1998. Ministers Announce Creation of Centre for Mine-Action Technologies, News Release, August 25, 1998. Available on line: *http://www.dnd.ca/eng/archive/1998/aug98/centremat_n_e.htm.*

Fredenburg, P.W. 1997. The banning of the antipersonnel landmine. Canadian Defence Quarterly 27(2): 5–9.

ICBL (International Campaign to Ban Landmines). 1999. Landmine Monitor Report 1999: Toward a Mine-Free World. Washington, D.C.: Human Rights Watch.

ICBL. 2000. Landmine Monitor Report 2000: Toward a Mine-Free World. Washington, D.C.: Human Rights Watch.

Metal Storm. 2000a. Landmine replacement a step closer to reality. Available on line: *http://www.metalstorm-ltd.com/press_releases/july03_00.html.*

Metal Storm. 2000b. Metal Storm secures Australian Army support for minefield replacement development programme. Available on line: *http://www.metalstorm-ltd.com/press_releases/july07_00.html.*

Roy, R.L., and S. Friesen. 1999. Historical Uses of Antipersonnel Landmines: Impact on Land Force Operations. Operational Research Division, Research Note 9906. Kingston, Ontario: Canadian Department of National Defence.

Appendix G

Mission Need Statements

BATTLEFIELD SHAPING AND FORCE PROTECTION AGAINST PERSONNEL THREATS[1]

Defense Planning Guidance Element

This requirement responds to the following:

- Defense Planning Guidance for FY99-03, page 84, 2 July 1997 requires the development of a resource plan addressing implementation of alternatives that meet the President's direction on Anti-Personnel Landmines (APL). Alternatives may be lethal, nonlethal, or a mix of the two, but will discriminate if they are lethal.
- Emerging Joint and Service Concepts of Operations (such as: Joint Vision 2010, Marine Corps Master Plan for the 21st Century, Army After Next, Global Engagement, etc.).
- Presidential Decision Directives 48 and 54, and Public Law 104-107 pertaining to the restricted use and banning of APL.
- Presidential statement of 17 September 1997 announcing the United States goals to end use of pure self-destructing APLs outside Korea by 2003. In Korea, alternatives are to be ready to replace pure APL (both non-self destructing and self-destructing) by 2006.

In response to both domestic and international efforts to restrict or ban the use of APL, the Department of Defense, with CINC and Service participation, has conducted assessments of APL utility as well as requirements for APL alternatives. Materiel and non-materiel alternatives to provide APL-like effects have been studied as part of the Office of the Secretary of Defense Working Integrated Process Team effort. Preliminary work showed technological alternatives that would have significant programmatic and operational implications over the 99-03 FYDP. Alternatives that protect our service men and women while addressing humanitarian concerns must be further explored and assessed. Alternatives must address CINC mission needs and be consistent with US policy and goals. The purpose of the MNS is to facilitate development of alternatives that address critical warfighting capabilities of battlefield shaping and force protection against personnel threats in full spectrum operations.

Mission and Threat Analyses

Mission

Military forces, operating in all environments and terrain, across the full spectrum of military operations, require capabilities for battlefield shaping and force protection that enhance operational and tactical flexibility. These capabilities must contribute to economy of force operations and provide force multiplier effects. Capabilities will enable deter, delay, and deny effects oriented on identified personnel threats using lethal and/or non-lethal means. These capabilities will enable friendly forces to:

- Delay, disrupt, and/or canalize enemy movement/ maneuver
- Deny enemy access to terrain or facilities (including short and long term deterrent for boundaries and DMZ areas)
- Enhancement of friendly force weapons, obstacles, and munitions effects (including Anti-Tank mines)
- Generate exploitable delays and opportunities (fix or contain enemy)
- Generate detection, alert, classification and/or early warning
- Produce desired effects on enemy forces (non-lethal to lethal)
- Reduce casualties/risk for U.S. and/or allied forces
- Deter pursuit to facilitate breaking of contact under pressure

[1] Landmine Alternatives Track III Broad Agency Announcement DAAE30-99-BAA-0103

The need exists to minimize risk to non-combatants while applying these capabilities to protect friendly forces as they shape the battlefield in a manner that recognizes economy of force concerns and force multiplier issues.

Threat Analysis

The DIA validated threat is documented in the: Land Threat Environment Projection Vol1 & 5, NGIC-1100-649-96, Feb 96: Threat to US Ground Maneuver Forces, Vol5: Special Operations Forces, NGIC-1100-653-97 Vol5, Nov 96. Statements of the threat are also documented in the: Marine Corps Intelligence Activity Mid-Range Threat Estimate 1997-2007; Army Land Warrior System Threat Assessment (LWSTA) dated 7 August 1994; Section I.C.6. of Defense Planning Guidance 1997-2001. Operations across the spectrum of conflict will expose US forces to threats ranging from highly organized to loosely structured groups of potentially hostile or adversary forces. Personnel threats will attempt to breach friendly force obstacles and barriers designed to deter, delay, or deny hostile forces.

The Threat can vary from large concentrations of infantry to localized numerically superior hostile personnel to small units operating covertly across the spectrum of military operations. Asymmetric or symmetric, unconventional or conventional threats, will operate in all terrain types and all environments to include urban areas and mega-cities, heavily forested/vegetated areas, and mountainous terrain.

Current Deficiencies - Shortfalls

United States APL policy directs the military to find alternatives for several fielded systems which are currently providing battlefield shaping and force protection capability against personnel threats. Deficiencies of the existing systems which must be corrected are:

- Current APL are target activated and are not designed to discriminate between combatants and non-combatants (no target classification, sensing, combat ID).
- Current non-self destructing APL needed for extended duration minefields are not designed to self destruct or self deactivate, therefore posing a residual hazard. Their target activation cannot be controlled (system effectiveness, capability, and availability).

Fielded landmine systems impose both finite capability and operational limitations/shortfalls on the military. Limitations/shortfalls refers to both enhancements to military value and humanitarian concerns. The following capabilities and limitations/shortfalls should be addressed:

- Current APL do not provide an interrupt capability between sense, warn and apply effect, nor do APL systems provide flexible command destruct options (command and control).
- Current APL apply only one level of force when activated (system capability).
- Hand emplaced minefields require a significant amount of logistical support and are manpower intensive; current self destruct APL systems are not recoverable or reusable (responsiveness, logistics).
- The obstacle or minefield is vulnerable to breaching (system effectiveness).
- Obstacles or minefields can limit friendly force mobility (control, combat ID) and pose fratricide risks.
- No extended-duration lethal obstacle capability exists to counter dismounted forces worldwide.
- Limited remote/autonomous ability for force protection.
- Limited ability to complicate and reconstitute obstacles.
- Current lethal pursuit deterrent / break in contact capability against personnel will not exist after 2003.

Timing and Priority

- Alternatives to Anti Personnel Landmines are high National priorities.
- An ADAM/RAAM conversion effort will be complete and in place by FY03.
- An APL-A full operational capability for USFK by FY 06 is required.
- The need exists for an APL-A outside of Korea.

Nonmateriel Alternatives

No feasible combinations of changes to doctrine, organization, concepts or training have been identified that satisfy the needs as constrained and defined in this MNS.

The OSD sponsored study conducted APL analysis at the tactical and operational levels. At both the Tactical and Operational levels, removal of APL from the combined arms synergy of combat effects has created an additional burden on the Tactics, Techniques and Procedures (TTP), organizational and equipment infrastructure and force structure. At the tactical level, modeling concluded, in most scenarios, that only significant increases in infantry or artillery can offset the loss of APL with corresponding increased losses of friendly forces. Additionally, Anti-handling Devices on Anti-Tank mines and Claymores (M18A1) are important and necessary military capabilities but are not effective as APL alternatives. Operational level modeling identified force structure alternatives that had the potential to perform APL functions, but required such forces (CAS, artillery, cavalry, attack helos, MLRS, etc.) to be in place and theater specific prior to commencement of hostilities. The loss of APL as a battlefield shaping and force protection asset will present a significant change to force ratios, force multipliers and tempo of operations.

Potential Materiel Alternatives

A variety of systems under development or available as off-the-shelf technologies and systems provide sensing, detection, delivery and effects that satisfy only portions of suitable capability alternatives. However, these alternative solutions impose force structure implications and are not easily adapted to satisfy all environmental, terrain and operational tempos. Furthermore, no system or simple system-of-systems has been identified which addresses the missions and deficiencies described previously. Additional analysis could occur in:

- C4ISR: Triggered detection devices (e.g., tripwire-like) not activating the munition until an external command to fire is received, this link could also provide command initiated self-dudding or destruction.
- Munitions: Replacement of current lethal munition payload with nonlethal or a variable nonlethal-to-lethal effects element or enhanced lethality munitions.
- Weapons: Joint Services Small Arms Program (Objective Individual Combat Weapon and Objective Crew Served Weapon, etc.).

Constraints

Overarching Constraints

The defined level of mission capability for all weather environments and under all terrain conditions across the full spectrum of conflict must satisfy an around-the-clock duty cycle. The following general constraints on the capabilities must be recognized:

- Able to provide target discrimination for lethal weapons systems in accordance with dynamic Rules of Engagement (ROE).
- Enable rapid operational and tactical deployment, employment, and recovery.
- Provide event- as well as time-based initiated actions.
- Possess Joint Technical Architecture (JTA) compliance for C4ISR interoperability.
- Comply with standards with regard to logistics, training, manpower, etc.
- Will not leave any greater residual hazards than non APL systems.
- Must be developed under aggressive material development timeline consistent with presidential policy.

Logistics

Weapons, munitions, and equipment with the capabilities to satisfy this MNS must be supportable within the existing Joint Service sustainment and maintenance concepts. The capability must possess Joint, Allied and coalition force interoperability (Joint C4ISR Interoperability Doctrine) and it must comply with transportation and employment standards with regard to logistics, training, and manpower.

Survivability

Solutions to this MNS will enhance and complement mission performance and will represent neither an operational encumberment nor a mission detractor. Requirements for operation, maintenance, or support will not increase fratricide, risk of attack or injury, or adversely impact on physical and mental fatigue. It is desired that the alternatives have increased effectiveness against hostile breaching or countermeasures. Components will have the same NBC survivability and decontamination survivability as currently fielded systems.

Operational Environment

1. Full Spectrum of Conflict; Major Theater War to Peace Operations.
2. Must include the capability to meet asymmetric threats where lethal force may not be desired. Scaleable to the degree of threat and scope of operations. Employment may be a system-of-systems, "layered defense" approach.
3. Urban and Mega-cities (MOUT).
4. Weather and Terrain. Solutions will operate and be maintained in all types of climate and terrain where US forces deploy.
5. Nuclear, Biological, and Chemical (NBC). Hardening to non-nuclear electromagnetic, radio interference and blast overpressure is required. Must be functional in an NBC environment while wearing NBC protective clothing, mission oriented protective posture IV gear.

MISSION NEED STATEMENT (MNS) FOR MIXED LANDMINE SYSTEMS ALTERNATIVES

Defense Planning Guidance Element

The purpose of this MNS is to facilitate development of alternatives to the Anti-Personnel (AP) submunitions in mixed landmine systems and/or to the entire mixed landmine system that addresses and responds to critical warfighting capabilities across the continuum of potential conflict. Alternatives must address both combatant Commanders in Chief (CINC) needs and be consistent with US policy goals and objectives. Current systems meet the combatant CINCs' requirements, and it is a warfighting imperative that gaps in capabilities not occur. This requirement responds to the following:

- Presidential Decision Directive/NSC-64 (PDD-64) of 23 June 1998. PDD-64 expands upon and strengthens US APL policy established in PDD-48 and PDD-54.

The PDD directs the Department of Defense to investigate the use of alternatives to existing Anti-Personnel Landmines (APLs) to replace the anti-personnel submunitions in mixed anti-tank mine systems and to actively explore the development of other technologies and/or tactical operational concepts which may result in a new, innovative approach to barrier systems that could replace the entire mixed munition; be advantageous militarily, cost effective, and safe; and eliminate the need for mines entirely.
- Defense Planning Guidance (DPG) for FY99-03 of 2 July 1997. The DPG requires development of a resource plan to implement the President's direction on anti-personnel landmines.
- Office of the Under Secretary of Defense for Policy Report to the Secretary of Defense on the Status of DOD's Implementation of the U.S. Policy on Anti-Personnel Landmines of May 1997. The report summarizes DOD direction to seek technological alternatives to APL and to review/modify operational doctrine, tactics and plans to reduce and ultimately eliminate reliance on APL.

Mission and Threat Analyses

Mission

Military forces, operating in all environments and terrain, across the full spectrum of military operations, require capabilities for battlefield shaping and force protection that enhance operational and tactical flexibility and set conditions for friendly dominant maneuver. These capabilities must contribute to economy of force operations and provide force multiplier effects against mounted and dismounted forces. They must degrade enemy capabilities and disrupt enemy maneuvers and operational tempo. Mission essential system capabilities include the ability to:

- Enhance the effects of close and deep friendly fires.
- Provide multiple methods of delivery.
- Provide a range of effects that inhibit mounted and dismounted maneuver.
- Resist the full spectrum of enemy breach methods including dismounted means.
- Provide early warning of a ground attack.

Additional warfighting needs include:

- Safety of use to own forces.
- Effectiveness in all types of terrain and weather.
- Minimization of residual hazard to own forces and non-combatants after military conflicts.
- Difficulty of detection by enemy forces.
- Minimization of fratricide.
- Selectable degree of effects against mounted and/or dismounted threat.
- Controllable activation/deactivation and duration before and after installation.
- Effectiveness in nuclear, chemical and biological environments.
- Ease and efficiency of distribution.

Threat Analysis

Operations across the spectrum of conflict will expose US forces to threats ranging from organized, conventional forces to loosely structured groups who fight symmetrically and asymmetrically. US forces will potentially be outnumbered and continue to face these threats in all terrain and environments. Enemy forces will equip themselves with increasingly advanced weapons and sophisticated countermine equipment. The dismounted soldier will continue to play a key role both in ground combat and in clearing the way for mounted forces irrespective of the type of opposing force. Military operations in areas with large populations of noncombatants (urban areas) are expected to increase in frequency of occurrence. The DIA validated threat is documented in the:

- Land Threat Environment Projection (LTEP) Volume 1: Threat to US Ground Maneuver Forces, NGIC-1100-649-96 Feb 96 and Volume 5: Special Operations Forces, NGIC-1100-649-97
- Marine Corps Expeditionary Warfare Threat Environment Projections (EWTEP) Executive Summary, MCIA-1234-001-98 and Volume 3: Direct Fire and Maneuver, Fire Support, and Engineering, MCIA-1143-001-98
- Marine Corps Intelligence Activity Mid-Range Threat Estimate 1997-2007
- Director Testimony to Senate Armed Services Committee on Worldwide Threat 1999
- Army Land Warrior System Threat Assessment Report (LWSTAR), 15 June 1998
- Agile Combat Support Threat Environment Description (TED), NAIC-1571-0664-98, July 1998.

Current Deficiencies - Shortfalls

While current systems meet the combatant CINCs' requirements, there are several system specific deficiencies that, if corrected, would enhance the warfighting capability. These deficiencies include:

- Susceptibility to increasingly sophisticated countermine methods.
- Reliance on pre-determined self destruct.
- Limited effectiveness in restricted and urban terrain.
- Logistically burdensome.
- Does not adequately contribute to situation awareness.
- No selectable target effect options.

A complete loss of current systems would create a greater set of deficiencies that would promote significantly higher levels of risk to US personnel and mission accomplishment. Regardless of region, nature of conflict or adversary, the capability afforded by our current systems must be retained as a minimum level of performance or must be enhanced.

Timing and Priority

The development of Mixed Landmine System Alternatives has been identified as a major DOD priority. Timing goals and objectives for the development of alternatives are outlined in PDD-64. Militarily advantageous, cost effective and safe alternatives must be fielded before ending the use of our current systems.

Nonmateriel Alternatives

No feasible combinations of changes to doctrine, organization, concepts or training have been identified that satisfy the needs defined in this MNS. OSD-sponsored studies at the tactical and operational levels show that elimination of mixed systems significantly increases risk and reduces effectiveness of the force. Tactical level modeling concluded that even significant increases in maneuver forces and supporting arms result in increased US combat losses and also cannot fully offset the loss of mixed systems. Additionally, M18 Claymore mines and anti-handling devices (AHD) on AT mines are important and necessary military capabilities but cannot substitute for mixed munitions. Operational level modeling identified force structure alternatives that had the potential to perform the battlefield functions of mixed systems but required a broad-spectrum of additional forces (tactical air, artillery, cavalry, attack helicopters, Multiple Launched Rocket System (MLRS), etc.) to be in place prior to commencement of hostilities. The loss of mixed system munitions as a battlefield shaping and force protection asset will present a significant negative change to force ratios, force multipliers and tempo of operations. No change in doctrine alone offsets the loss of mixed systems without associated changes in force structure. Replacement of mixed system munitions by other types of obstacles is not as effective, mandates an increase and reallocation of men and equipment, and detracts from other missions.

Potential Materiel Alternatives

The variety of systems fielded, under development or available as off-the-shelf technologies satisfy only portions of the required capabilities. In addition, these alternative solutions have force structure implications and are not easily adapted to satisfy all operational tempos and weather and terrain conditions. Since no system or system-of-systems has yet been identified that fully addresses the missions and deficiencies described previously, additional analysis is required to determine how critical technologies can be effectively integrated into a robust system. These critical technologies include sensors and target recognition, command and control, nonlethal and lethal munitions and effects, and communication/low power networks. Emphasis should be placed on these system characteristics: tunable munitions effects, countermeasure resistance, all terrain and weather operation, target identification and tracking, early warning, and situational awareness.

Constraints

Key Overarching Constraints

The new system must have the ability to support the full spectrum (or range) of military operations from peace operations to full military conflict. It should be designed to address dismounted, mounted, or a combination of these two types of threat maneuver methods. The new system must be all weather capable during delivery/emplacement and after activation. It should be able to be deployed and operate in all types of terrain and vegetation, to include steep gradient mountains, urban areas and the littorals. The new system should be effective on hard road surfaces independent of the delivery method. The new system design should be easily incorporated into inter-service tactics, techniques, and procedures utilizing current platforms where possible, and interoperable with other coalition and allied military forces. Communications and computer equipment should be designed to be compliant with the Defense Information Infrastructure Common Operating Environment (DII COE), Joint Technical Architecture (JTA) as well as interoperable with existing U.S. forces inventory [C4ISR], and allied forces. The new system should address all MANPRINT domains to enable rapid, safe and reliable deployment of the equipment. The system should be able to be employed throughout the theater of operation either by hand or utilizing remote air delivery methods. The design should allow the system to operate without degradation in a biological/chemical environment, and should have the same operational survivability as similar existing systems in a nuclear environment. The design must allow for the system to be removed or rendered inoperable once the tactical situation requires that the new system's effects be removed from the battlefield. If the system is destroyed, it must minimize residual hazards. The system or its sub-components must not be controllable or reuseable by belligerents at any time.

Logistics

The maintenance concept for the new design should incorporate common troubleshooting and repair procedures and should not require new military occupational skills or additional manpower to operate or maintain the system. The system should be able to be configured for transportation in

U.S. and allied countries using common light, medium, and heavy ground haul assets as well as military air and sea platforms. The total system design should incorporate minimum packaging materials and be packaged so that the new system can be stackable, storable, and easily inspected throughout its shelf life. Packaging for an individual system intended for hand-emplacement should incorporate MANPRINT design features focusing on lightweight characteristics.

Survivability

The new system should be capable of airborne and remote delivery conditions in all expected operational environments and conditions. The system should be re-useable if not fully deployed, and the design should incorporate all MANPRINT domains to ensure the system is durable. The new design should make the system resistant to common battlefield effects, including small arms (up to .50 cal), small fragmentary (hand grenade), fire, overpressure, and radio frequency signals. The new system should be recoverable, corrosion resistant, electro-magnetic pulse hardened, and remain functional after biological or chemical attack contamination.

Operational Environment

The new system must be insensitive to the wide range of weather effects where U.S. forces deploy, both during the delivery/emplacement and after activation of the system, to include effects from wind, lightning, fog, rain, snow and precipitation in general. Once activated, the system must operate in the residual and time-based effects of weather to include build up of snow, ice, frost, and mud to provide an around-the-clock duty cycle operational capability. If sensors are incorporated in the new design, they must be able to operate in extreme hot and cold environments and should be insensitive to temperature cycles associated with diurnal temperature variations.